CAMBRIDGE COUNTY GEOGRAPHIES

SCOTLAND

General Editor: W. Murison, M.A.

RENFREWSHIRE

Cambridge County Geographies

RENFREWSHIRE

by

FREDERICK MORT

M.A., B.Sc., F.G.S., F.R.S.G.S.

With Maps, Diagrams and Illustrations

Cambridge:

at the University Press

1912

CAMBRIDGE UNIVERSITY PRESS
Cambridge, New York, Melbourne, Madrid, Cape Town,
Singapore, São Paulo, Delhi, Mexico City

Cambridge University Press
The Edinburgh Building, Cambridge CB2 8RU, UK

Published in the United States of America by Cambridge University Press, New York

www.cambridge.org
Information on this title: www.cambridge.org/9781107616509

First published 1912
First paperback edition 2013

A catalogue record for this publication is available from the British Library

ISBN 978-1-107-61650-9 Paperback

CONTENTS

CONTENTS

ILLUSTRATIONS

ILLUSTRATIONS

MAPS AND DIAGRAMS

The illustrations on pp. 5, 29, 31, 32, and 87 are from photographs by Mr J. W. Reoch; that on p. 38 is from a photograph by Mr J. Frew; that on p. 115 is reproduced by kind permission of the Society of Antiquaries of Scotland; that on p. 156 is from a photo by Messrs T. & R. Annan; that on p. 76 from a photo by Mr Robertson of Greenock; and the remainder are from photos by Messrs J. Valentine & Sons.

1. County and Shire. The Origin of Renfrewshire.

The modern county is a political unit. It is the division of a kingdom administered by a sheriff, and this system dates back at least as far as the reign of David I (1124–1153). All such divisions may be called *counties*, but it is only to some of them that the name *shire* can be applied. Caithness and Sutherland, for example, are counties, but not shires, while Renfrew may be called either shire or county. An explanation of the names makes this distinction clear. The word *shire* is said to be allied to *share* and *shear*, and consequently to the Anglo-Saxon *sceran*—to cut. It would therefore mean a piece cut off. Professor Skeat, however, now derives *shire* from Anglo-Saxon *scirian*—to distribute, appoint. The shires were portions of a kingdom which were originally governed by the great earls of the country, who in many cases took their titles from the districts they ruled. Renfrewshire was a part of the old kingdom of Strathclyde. When William I had conquered England, many of the English earls were dispossessed of their lands, which were given to William's companions or *comites*. Each district was

therefore called a *comitatus*, or, in its French form, *comté*,
from which we get the word *county*. The counties of
Caithness and Sutherland were in the hands of the Scan-
dinavian Earls of Orkney (whence the name Southern
Land), until the very end of the twelfth century, when
they were subdued by William the Lion. When they
became attached to the kingdom of Scotland the Norman
terms were already in use, and therefore the Anglo-Saxon
name does not apply to them.

Although the counties are divisions administered by
the sheriffs of a king, their evolution has been a complex
process. They are the final results of a long series of
adjustments between different forces. The king, the
church, the nobles, and in modern times the burghs,
were centres of segregation that tended to group the
community in different ways. Thus it happens that
there is still a considerable amount of overlapping and
confusion in the administrative divisions, not only of
Renfrewshire, but of all the counties of Scotland. Yet
working through all these discordant forces, the geographical factor is visible. The physical characteristics
of a district have directed the other forces, and moulded
the political divisions in harmony with natural regions.
Of this fact Renfrewshire is a good example. It is hardly
so complete a geographical unit as Lanarkshire, which
comprises simply the upper and middle Clyde basin, but
its boundaries have a well-defined geographical basis.
The point at which a large river becomes too wide to
be bridged is of prime importance. The stream of traffic
down the valley divides here, and the up-river trade

coalesces at this point. Hence a large town often grows up at such a place, and, at this place also, counties often terminate. Such is the case with the Clyde. Lanarkshire ends just where the river becomes too wide to be bridged conveniently. Above this point the banks of the river are embraced by one county. Below it, the river forms the boundary between Renfrew and Dumbarton.

The southern boundary is also in the main a natural one. A broad ridge of flat-topped, volcanic hills runs from south-east to north-west, separating Lanarkshire from Ayrshire. A continuation of the same ridge separates the latter county from Renfrewshire. The ridge is broken through by the Loch Libo valley and by the Lochwinnoch valley, but it keeps on its course and reaches its highest point in Hill of Stake on the borders of Renfrew and Ayr. The eastern boundary is a compromise between Lanark and Renfrew, in other words, a line approximately separating the middle from the lower basin of the Clyde.

Originally there was no such separation. Lanarkshire and Renfrewshire were one. William Hamilton of Wishaw, writing about 1710, tells us that "The shyre of Lanark was anciently of greater extent than now it is; for there was comprehended in it the whole sheriffdome of Ranfrew, lying laigher upon Clyde...untill it was disjoyned therefra by King Robert the Third, in anno 1402." Since then the changes in the boundaries of Renfrewshire have been geographically of little moment. Twenty years ago the Boundary Commissioners transferred certain areas from one parish to another, in some instances from one county to another, in order to rectify anomalies of

administration, but these changes were not of great importance save from the administrative point of view.   The
name Renfrew is said to be derived from *rhyn*—a point
of land, and *frew*—the flowing of water; there are, however, other explanations of the word.   The district was
formerly called Strathgryfe from the name of one of its
most important rivers.

2. General Characteristics — Position and Relations.

Of all the counties of the west of Scotland, there is
only one that is entirely within the Lowlands, and this
one is Renfrew.   By this it is not meant that the whole
shire is low-lying.   Far from it.   A large proportion of
the total area is hilly, but the hills are of the "Lowland"
type.   This paradox requires further explanation.

There are in Scotland three well-marked natural
divisions, the Highlands, the Central Lowlands, and the
Southern Uplands; and these three districts differ strongly
in physical aspect, in rocks, in scenery, in vegetation, and
in industries.   The Central Lowlands are separated from
the Highlands by a line running from north-east to south-
west, between Stonehaven and Helensburgh; they are
separated from the Southern Uplands by an almost parallel
line, running from St Abb's Head to Girvan.   Dumbarton,
Lanark, Ayr, Bute, all are crossed by one or other of these
two lines; Renfrew alone falls entirely between them.
These lines mark the course of two great faults or cracks,

which traverse the whole country, and between which the land has gradually sunk for thousands of feet. This sinking of the central part of Scotland took place many ages before man appeared on the earth ; yet it may be considered the most important stage in the evolution of Scotland, for it preserved the all-important coalfields of

Types of Hills: Highland Type (Loch Long and Loch Goil)

the Lowlands, on which the prosperity of the country largely depends. The Central Lowlands of Scotland are not only the most fertile part of the country, but by far the greatest proportion of the mining and the manufactures is carried on there. It has thus become a district unique in Great Britain, for it possesses the characteristics of at least three separate parts of England—the south-eastern plain devoted to agriculture, the "Black Country" with its coal and iron industries, and Lancashire with its cotton manufactures.

The hills of the Central Lowlands—Sidlaws, Ochils, Pentlands, Campsies, Kilpatricks, Kilbarchans—are all of a similar kind (the "Lowland" type above referred to), and differ markedly both from those of the Highlands,

Types of Hills : Southern Upland Type (the Lowthers)

and from those of the Southern Uplands. As will be shown more fully in the section on Geology, this difference is due to the nature of the rocks. From an elevated spot within the Highlands there is revealed on all sides a bewildering chaos of mountain and valley. As a rule this tumbled sea of peaks rises into bare rock, sometimes rugged, splintered, and pinnacled, sometimes upheaving a huge,

rounded shoulder of rock terminating in an abrupt pre-
cipice. In the Southern Uplands the wildness, ruggedness,
and grandeur of the Highlands as a rule are absent, for the
outlines of the hills are generally smooth and rounded;
yet there is a pure and softly flowing sweep of contour,
and a charm of delicate colour about these green and
treeless summits, found nowhere else in Scotland. The

Types of Hills: Lowland Type (looking across
Castle Semple Loch)

Lowland hills consist of large, irregular masses of volcanic
rocks. They are table-lands, something less than 2000
feet above sea-level, with undulating surfaces, rising into
no prominent peaks, and thus differing from both the
other types. As a rule the sides of these hills rise very
steeply from the low ground, but once the top is gained,

one can walk for many miles over bare moorland, the
surface rising or falling within the limits of a hundred
feet or so.   All the hills of Renfrewshire are of this
kind.

The position of Renfrewshire on the western slope
of Scotland was at first a disadvantage.   For a long time
the eastern coastal plain was by far the most important
part of the country.   The commerce of Europe to a large
extent was carried on in the districts bordering the North
Sea.   The face of the county was thus turned away from
the chief commercial centres; but the progress of civilisation in its westward march, particularly the development
of American trade, has shifted the centre of gravity of
commerce to the shores of the Atlantic, and thus the
geographical position of Renfrewshire at the present time
is one of its most important advantages.

Although the mineral wealth of Renfrew is not of
great value, the county shared in the phenomenal growth
of Scottish industrial centres during the last century and
a half.   It is near enough the rich coalfields of Lanarkshire to participate in the prosperity that came with their
development.   The banks of the Clyde with their easy
access to the sea, and their proximity to valuable coal and
iron fields, formed an ideal home for the ship-building industry, and in this branch of trade Renfrewshire is an easy
first, claiming yearly about half the total tonnage of the
river.   Yet it must not be forgotten that of itself the
county could not have attained this industrial eminence.
It must obtain its supplies of coal, iron, and steel elsewhere, but this consideration only serves to throw into

stronger relief the energy, the skill, and the enterprise that brought the county to the front in spite of the disadvantages it laboured under in its lack of mineral wealth.

But Renfrew is by no means purely a manufacturing county. Its pursuits are numerous and well-balanced. Agriculture and shipping complete the trinity of its most important interests. The county falls naturally into two main divisions, hilly and low-lying, and the former is of no industrial importance. The low ground may be sub-divided again into three parts, a broad flat area in the east, submerged below the waters of the Clyde in comparatively recent times; a narrow coastal strip bordering the hills on the north and west (an old sea beach in fact), and lastly the open valleys of the interior. It makes for clearness to think of the first of these districts as the home of manu-factures, the second of the shipping trade, the third of agriculture. Of course in nature, divisions are never absolutely sharp; and thus manufactures and agriculture are found to some extent over all the lowlands of the shire.

3. Size of County. Boundaries.

Although in many respects Renfrewshire can more than hold its own among the counties of Scotland, in size it takes a very humble place. It ranks twenty-seventh among the thirty-three counties of Scotland. The largest shire, Inverness, is more than seventeen times the size of

Renfrew; and yet it is a striking fact that Inverness contains only about one-third the population of the smaller county. Renfrewshire is an irregular oblong, the longer axis of which lies roughly north-west and south-east. From Cloch Point to the south-east extremity of the county between Carse Hill and Laird's Seat the length is a little over 30 miles, while the greatest breadth from Kilbirnie Loch to Erskine Ferry is over 13 miles. The total area, including foreshore and inland water, is about 250 square miles.

In the main the boundaries are simple and formed by important geographical features. Roughly speaking, the northern boundary is the Clyde, the eastern is the White Cart, the southern is the watershed of the volcanic hills that run from south-east to north-west, and the western boundary is the Firth of Clyde. We shall next trace the boundary in some detail beginning at Greenock. From that town to Yoker Burn the boundary is the natural and obvious one formed by the broad waters of the Clyde. Then strange to say the boundary runs up the Yoker Burn *north* of the Clyde, passes east to Temple, turns south through Anniesland and along Crow Road to Jordanhill Station, then south-west to the Clyde again at the old Marline Ford. This little, detached part of the county, left stranded on the north side of the Clyde, will later be discussed in more detail. Crossing to the south side of the Clyde the boundary strikes west along an old channel of the river. On reaching the outskirts of Renfrew the line doubles back sharply, and runs south-east, passing just to the south of Craigton Cemetery.

Thence it runs to Pollokshaws, where it meets the White Cart, which it follows for a mile. Next it swings round Cathcart, passing over a mile to the east of the village, and then bends west till it rejoins the river Cart a mile south of Cathcart. It follows the course of that stream to its junction with the Threepland Burn, whence it runs

Wemyss Bay

due south to the crest of the hills where Renfrew, Ayr, and Lanark meet. It would be tedious to trace in great detail its north-west course from here to Caldwell. It is enough to state that it runs along the tops of the moors, keeping just a little south of the actual watershed. From Caldwell the boundary takes a sinuous course to the north end of Kilbirnie Loch, and then mounts the hills again.

The highest points of the shire are reached in Misty Law and the Hill of Stake, a drainage centre whence the streams radiate out in all directions. From the Calder Water the boundary passes to a feeder of Loch Thom and thence to the Kelly Burn, down which it runs to the pier at Wemyss Bay. From Wemyss Bay back to Greenock the county is bounded by the salt waters of the firth.

To sum up, the detailed tracing of the boundary has shown that the line is not an arbitrary one. Nature has forced man to draw his lines of demarcation in accordance with her dictates. Mountains and broad rivers are every-where natural barriers, and even such minor separation lines as those of counties are, as we have seen, powerfully influenced by physical features. Thus the tracing of the boundaries of the shire has afforded us an excellent illus-tration of geographical control.

At one point, however, the geographical control seems to be set at defiance. The boundary crosses the Clyde in order to include Jordanhill and the surrounding districts, which are thus quite detached from the rest of the county. The attention of the geographer is at once arrested by this remarkable circumstance, although historians have not thought it worthy of discussion. The explanation seems to be partly geographical, partly historical. If the Clyde had always been the deep river that it now is below Glasgow, it would have formed such a formidable barrier to cross-communication that the boundary of the sheriff-dom would probably never have overstepped it. But before the Clyde was deepened, the appearance of the river near Renfrew was quite different from what it now is. Even in James Watt's day there were a dozen shoals between Renfrew and Glasgow, and in addition the river formerly split into two shallow branches. In ordinary weather, therefore, communication between both banks was easy; in other words the geographical control exerted by the river was weak. Again, historically considered, the lands north of the Clyde formed part of the ancient barony of Renfrew, for, as we shall see later, they had been given by the king with other parts of the county to Walter the High Steward. It was the same reason that made Bathgate and the surrounding territory part of the sheriffdom of Renfrew until the sixteenth century, when the office was sold by the second Lord Semple. This district was obtained by Walter the sixth Steward, son-in-law of Robert the Bruce, as part of the dowry brought him by Marjory. In legal usage the district is still referred to as a sheriffdom separate from Linlithgow.

4. Surface and General Features.

The surface of Renfrewshire is extremely varied. It ranges from sea-level to a height of over 1700 feet. The loftiest part of the county is the southern boundary, which forms a high rim to the shire from which the land falls to the north-east, sometimes gradually, sometimes very steeply. The hill masses have the structure of plateaus. There is nothing in the least approximating to a *range* of hills. To endeavour to represent the hills on a map by the favourite devices of lines, or "herring-bones," would be ludicrously

inexact.   They must be shown as broad, irregular areas,
as is done in the map on the front cover, or more simply
in the sketch-map on p. 140.   In addition there are a few
steep, isolated crags which will be referred to more particularly in the section on Geology.

From the south-eastern extremity of the county a broad
band of high moorland stretches without a break to the
Firth of Clyde, save where it is deeply cut by two important valleys trenching at right angles across the general
direction of the uplands.   These two valleys, therefore,
divide the hill masses into three blocks.   That to the
south-east comprises the Eaglesham, Mearns, and Neilston
moors.   The middle mass stretches from the Loch Libo
valley to the valley of the Black Cart.   Unfortunately
there is no name for this hill mass as a whole.   In default
of a better, the name Corkindale Moors (from Corkindale
Law, the highest point) will here be adopted.   The Braes
of Gleniffer, the northern edge of these hills, have been
rendered classic by Tannahill.   The third plateau stretches
from the Black Cart valley to the Firth of Clyde.   Only
the northern part of these hills is in Renfrewshire.   They
stretch north and south without a break for nearly 20
miles, from Greenock to Ardrossan.   There is no generally accepted name for these hills, but many years ago
James Geikie called them the Kilbarchan Hills, and that
name will be retained here.

The hill masses of Renfrewshire are flat-topped on the
whole, and therefore we find neither the imposing peaks,
nor the sharp, serrated ridges that give the note of grandeur
to a typical Highland scene.   Occasionally, however, the

The Linn, Gleniffer

hills rise into distinct summits, the highest being Hill of
Stake (1711 feet), East Girt Hill (1673 feet), and Misty
Law (1663 feet), all near each other on the central and
culminating part of the Kilbarchan Hills.   This district
is one of the most unfrequented in the west of Scotland.
The tide of railway traffic pours down the Lochwinnoch
valley, and swinging round the south of the Kilbarchan
Hills flows north again to Largs; but it leaves these lonely
moors untouched.   They make no arresting appeal to the
eye that is anxiously on the alert to catch the first glimpse
of the glorious, splintered peaks of Arran.   They are bleak,
lonely, and treeless, clothed with heath or coarse bent, with
bare, rocky ribs protruding here and there.   And yet they
have a nameless, powerful fascination of their own.   The
air is clear and pure, and the miles of undulating walking
afford a sense of freedom and an impression of space that
true peak-climbing never gives.   But above all, the views
from the hills of Renfrewshire are unsurpassed in Scotland.
From Misty Law, or even Corkindale Law farther inland,
a vast panorama is unrolled on a clear day.   Northwards
lies the broad valley of the Clyde, a dark pall of smoke
marking where Glasgow blackens the landscape.   Here
and there a gleam of silver betrays the course of the river
as we sweep the horizon from Dumbarton Rock to Tinto.
To the north-west, blue in the distance, are the rugged
peaks that form the outposts of the giant armies of the
Highlands, the massive bulk of Ben Lomond, the gashed
outline of the Cobbler, and the shapely cone of Ben Ime,
prominent among less noticeable mountains.   Westward
the eye lingers on the jagged, granite peaks of Arran, then

passes to the blue form of Ailsa Craig, seeming to hang like a tiny hay-stack in mid air. Perhaps even a glimpse of Skiddaw may be had in the far south, if the air has been recently washed by rain, and is unusually clear and free from dust. Although the view from the high interior hills is the most extensive to be obtained in Renfrewshire, it is not the finest. For that we need climb only to a modest elevation on any of the hills behind Gourock or Greenock. Then to the most of the features mentioned above must be added a near prospect of the blue waters of the Firth, sparkling in the sunshine, or lying one mass

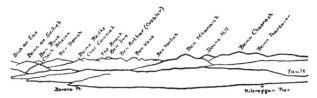

Southern Highlands as seen from behind Gourock

of molten gold and crimson, as the sun sinks behind the mountains of Cowal. It is one of the finest views in Britain. From such a point a splendid panorama may be had of the southern rampart of the dissected plateau that we call the Highlands. It is by no means easy to identify the individual summits without instrumental aid, therefore a drawing is given (see above) based on photographs and theodolite readings, showing the most interesting part of the mountain panorama seen from the hills behind Gourock.

It has been stated that the hills of Renfrewshire are

divided into three main masses by two large valleys which run north-east and south-west. These valleys are of the utmost importance, for they form the only easy routes from Glasgow and the large towns of the lower Clyde to the busy and fertile Ayrshire plain. The larger of these two valleys is that which runs from Johnstone to Dalry; the other begins at Barrhead, and passes through Caldwell and Lugton. Such valleys breaking through hilly barriers and forming easy routes from one rich district to another are often called "gaps," and these valleys may be termed respectively the Lochwinnoch Gap and the Loch Libo Gap. Ebbing and flowing through these natural gateways, pours a ceaseless tide of traffic by road and rail. The main line to Ayr and Stranraer goes through one; the main line to Kilmarnock, Dumfries, and London uses the other. These valleys are as interesting in origin as they are important for trade. They are of that peculiar class known as "rift valleys," of which the best known type on a large scale is the Great Rift Valley of Africa that runs from Lake Nyasa to Palestine. The crust of the earth has been fractured along two parallel lines, and the strip between has sunk, leaving a wide valley-floor, bordered by steep sides. The Red Sea is a part of the African rift, so is the curious Jordan valley, in places 2000 feet below sea-level. Its course is marked by narrow lakes—Nyasa, Tanganyika, Albert, Rudolf, Dead Sea, Galilee. The Loch Libo Gap and the Lochwinnoch Gap are similar, though on a smaller scale. Loch Libo lies in the former; Castle Semple Loch, Barr Loch (now drained), and Kilbirnie Loch mark the course

of the latter.   All these lakes were formerly more extensive.   In fact even yet in times of flood the Lochwinnoch
Gap is covered by an almost continuous sheet of water
that stretches from Howwood nearly to Dalry.   These
valleys are green and pleasant places, sheltered and fertile,
while the broad lakes of still water with their background
of hills give the scene a quiet charm.   It is the type of
scenery that most impressed the older writers.   Of Loch
Libo the parochial enthusiasm of the writer of *The New
Statistical Account* led him to state that "Loch Libo presents a scene of unparalleled beauty," and even to maintain that "Loch Libo excells in picturesque scenery Rydal
Water in Cumberland."   The beauty of the loch and its
surroundings, especially towards the close of a still summer
day, is undeniable, but to compare it with the Lake District
is rash.

The north-eastern part of the shire is extremely flat.
This district, known as the "laigh lands," was at no very
remote epoch covered by the waters of the sea.   The sites
of busy industrial centres such as Johnstone, Paisley, and
Glasgow were occupied by the wide estuary of the river,
on the surface of which primitive man propelled his rude
dug-out canoe.

That much of what is now dry land must have been
under water in human times is shown by the canoes that
have been found at various places in the area.   In Roman
times, however, the relation between land and water was
what it now is, for the position of Antonine's Wall shows
that there has been no appreciable rise or fall of the land
since it was made.

5. Watershed. Rivers. Lakes.

We have already seen that the southern boundary of the county coincides approximately with the watershed. From the south-eastern portion of the shire to the Loch-winnoch Gap the watershed is slightly north of the boundary, and thus a narrow strip of the county drains into Ayrshire. From the Hill of Stake streams radiate out in all directions. The moors north of the Hill of Stake form a divide between the short streams flowing to the Firth, and the innumerable burns draining to the east, chiefly into the Gryfe. There is a prevalent but mistaken belief that a watershed must be a range of hills, or at any rate must stand well above the level of the surrounding country. In many cases this is not so. The watershed may be a flat marsh, and one may sometimes walk right across an important watershed without noticing any change of slope whatsoever. This is illustrated by the watershed in the two chief Renfrewshire valleys. In the Loch Libo Gap the divide occurs between Neilston and Caldwell, but it would be difficult to say exactly where, for the ground is apparently level. The case is even more striking in the Lochwinnoch Gap, where the watershed practically disappears; for Castle Semple Loch draining north, and Kilbirnie Loch draining south are frequently connected by tiny natural canals. It is an excellent illustration in miniature of what we find on a gigantic scale in South America, where the basins of the Orinoco and the Amazon are connected by the Cassi-quiare.

Practically the whole shire is drained by three streams and their tributaries, the Gryfe, the Black Cart, and the White Cart.   The Black Cart flows from Castle Semple Loch, and in parts of its course resembles a canal as much as a river.   It is a sluggish stream that falls less than 90 feet in a length of 10 miles.   The Gryfe rises on Creuch Hill and flows through the Gryfe Reservoir.   In

The Cart at Busby

its upper course it traverses the bleak uplands that form the northern part of the Kilbarchan Hills.   Lower down, Strathgryfe is a pleasant vale, fertile and in parts richly wooded.   There is good trout fishing to be had, although the waters are preserved.   The White Cart is the classic river of the county.   It rises among the moors of Eaglesham near the junction of Ayr, Renfrew, and Lanark, and for several miles it forms the boundary between the two latter counties. It flows through Busby past the print-works that now stand empty and silent, mementoes of the days when trade was better. Northwards still the stream flows past the old castle of Cathcart, the origin of which is lost in the mists of time, but which was a strong place in the days of Wallace. Camphill forces the river to swing aside to the west through the policies of Pollok and past the old castle of Crookston to Paisley. As it enters the town it turns north again, and after receiving the Black Cart on its left bank it joins the Clyde opposite Clydebank.

In its upper and middle reaches the scenery of the White Cart is almost invariably pleasing, and in places is even beautiful. The banks are richly wooded, and fine mansions are dotted here and there along the valley. It has been a very home of the muses, and has inspired perhaps more bards than any other stream in Scotland. Its praises have been sung by Burns and Tannahill. Upon its banks Thomas Campbell spent the summers of his youthful years, and retained tender memories of these till his latest days—

"O scenes of my childhood, and dear to my heart,
   Ye green waving woods on the margin of Cart,
   How blest in the morning of life I have strayed
   By the stream of the vale and the grass-cover'd glade."

Christopher North received his early education in this locality; and in *A Summer in Skye*, Alexander Smith has given us a sketch of this district, one of his boyhood's

Rouken Glen: The Falls

haunts. Pennant declared that at the junction of the
Cart and the Gryfe the scenery was "the most elegant
and softest of any in North Britain." But, alas, here as
elsewhere, man strives his utmost to pollute the fair river
with pestilential sewage and every form of industrial waste.
But in this respect a better day is dawning, thanks to the

Lovers' Walk, Rouken Glen

enterprise of Glasgow, which shines the brighter against
the black background of indifference shown by some
Renfrewshire communities.

Some of the smaller valleys are of considerable interest
and beauty. Devols Glen near Port Glasgow is not only
charming in itself, but possesses a waterfall that will bear

comparison with many more widely known. The Waulkmill Glen near Barrhead, and the Rouken Glen in Glasgow's new pleasure ground, are as pretty bits of stream scenery as the heart of man could desire, and their charm is enhanced by their nearness to large industrial centres.

We are apt to forget that for a short distance—between Whiteinch and Yoker—the Clyde is purely a Renfrewshire river. But how different from the days of old, before human skill and perseverance had changed the very face of nature! It is exactly the Renfrewshire part of the river that has been altered most. Before Glasgow had transformed it, the Clyde here was broad and shallow, and split up into several branches. In the middle of the seventeenth century there were eight islands between the Kelvin and Erskine Ferry. The map on p. 26 gives a good idea of the appearance of the river at this time. It is from a map published at Amsterdam in 1654, and reprinted in Crawfurd's *History of Renfrewshire*. "Whyt inch" is now part of the north bank of the river, and King's inch is no longer separated from the royal burgh by a branch of the Clyde, although both districts retain their old names of *inch* or island. The river is now half a mile distant from Renfrew, yet the title-deeds of some of the houses still give the Clyde as the boundary of their gardens.

Lakes are numerous in the upland parts of the shire, many of them being utilised as reservoirs for the industrial centres of the lowlands. Lakes in fact are a characteristic feature of the Renfrewshire type of hills. As we have seen, the hills are flat-topped with an undulating

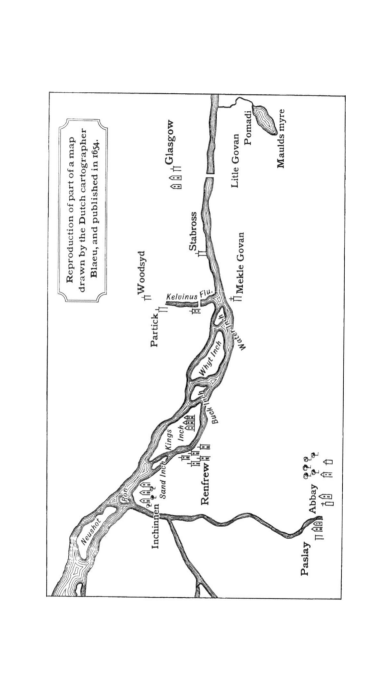

Reproduction of part of a map drawn by the Dutch cartographer Blaeu, and published in 1654.

surface, and thus the hollows of the table-land form natural receptacles for collecting the drainage of a fairly extensive area. Dotted here and there all over the hills of the county they occur, and their presence renders pleasing a type of scenery that otherwise would be bleak and monotonous. Loch Thom and the Gryfe Reservoirs form a charming little group of lakes within easy reach of Greenock and Port Glasgow. The Balgray Reservoir near Barrhead belongs to the Corporation of Glasgow, and with its beautifully kept surroundings, overlooked by the romantic towers of Pollok Castle, it makes a charming picture. It would be tedious to enumerate all the lochs dotted everywhere over the hilly portions of the shire. Their positions and names can best be seen from the map on the front cover.

The valley lakes present some special points of interest. We have seen that Loch Libo and Castle Semple Loch occur in peculiar valleys of the rift type; long, narrow lakes, indeed, being characteristic of such valleys. These lochs undoubtedly occupied a much greater extent in former times than they do now. South of Castle Semple Loch there was formerly another lake called Barr Loch, which was drained in the eighteenth century. Castle Semple Loch is now a mile and three-quarters long and half a mile broad, but Hamilton of Wishaw writing 200 years ago states that it was then three miles long, while Principal Dunlop writing a little earlier gives the dimensions of "Lochquhinnoch" as four miles by two miles. Again in the seventeenth century map of Renfrewshire already referred to we find "Loch Whinnoch" and the

"Loch of Kilbyrny" connected by a narrow neck of water. But it is useless to pile evidence on evidence for what is a fairly obvious fact. It need only be added that in flood-time we get a good idea of the appearance of the valley in former times. For many miles there stretches a long sheet of water, continuous save where a road or a railway embankment cuts across the temporary lake like the dykes of Holland. The original great lake probably drained into the sea at Irvine. But when the Black Cart succeeded in cutting through the rock barrier near Elliston, the lake waters would be carried off to the Clyde, the level would be lowered, and previously submerged river deltas would then divide the original sheet of water into a chain of lakes.

6. Geology and Soil.

The rocks are the earliest history books that we have. To those who understand them they tell a fascinating story of the climate, the natural surroundings, and the life, of a time many millions of years before the foot of man ever trod this globe. They tell of a long succession of strange forms of life, appearing, dominating the world, then vanishing for ever. Yet not without result, for each successive race was higher in the scale of life than those that went before, till man appeared and struggled into the mastery of the world.

The most important group of rocks is that known as *sedimentary*, for they were laid down as sediments under water. On the shores of the sea at the present time we find accumulations of gravel, sand, and mud. In the course of time, by pressure and other causes, these deposits will be consolidated into hard rocks, known as conglomerates, sandstones, and shales. Far out from shore there is going on a continual rain of the tiny calcareous

Types of Rocks—Sedimentary
(*Giffnock Sandstones*)

skeletons of minute sea-animals, which accumulate in a thick ooze on the sea-floor. In time this ooze will harden into a limestone. Thus by watching the processes at work in the world to-day, we conclude that the hard rocks that now form the solid land were once soft, unconsolidated deposits on the sea-floor. The sedimentary rocks can generally be recognised easily by their bedded

appearance. They are arranged in layers or bands, sometimes in their original, horizontal position, but more often
tilted to a greater or less extent by subsequent movement
in the crust of the earth.

We cannot tell definitely how long it is since any
special series of rocks was deposited. But we can say
with certainty that one series is older or younger than
another. If any group of rocks lies on the top of another,
then it must have been deposited later, that is, it is
younger. Occasionally indeed the rocks have been tilted
on end or bent to such an extent that this test fails, and
then we must have recourse to another and even more
important way of finding the relative age of a formation.
The remains of animals and plants, known as fossils, are
found entombed among the rocks, giving us, as it were,
samples of the living organisms that flourished when the
rocks were being deposited. Now it has been found that
throughout the world the succession of life has been
roughly the same, and palaeontologists (students of fossils)
can tell, by the nature of the fossils obtained, what is the
relative age of the rocks containing them. This is of
very great practical importance, for a single fossil in an
unknown country may determine, for example, that coal
is likely to be found, or perhaps that it is utterly useless
to dig for coal.

There is another important class of rocks known as
igneous rocks. At the present time we hear reports at
intervals of volcanoes becoming active and pouring forth
floods of lava. When the lava solidifies it becomes an
igneous rock, and many of the igneous rocks of this

country have undoubtedly been poured out from volcanoes that were active many years ago. In addition there are igneous rocks—granite for example—that never flowed over the surface of the earth in molten streams, but

Types of Rocks—Igneous

solidified deep down in subterranean recesses, and only became visible when in the lapse of time the rocks above them were worn away. Igneous rocks can generally be recognised by the absence of stratification or bedding.

Sometimes the original nature of the rocks may be

altered entirely by subsequent forces acting upon them. Great heat may develop new minerals, and change the appearance of the rocks, or mud-stones may be compressed into hard slates, or the rocks may be folded and twisted

Types of Rocks—Metamorphic

in the most marvellous manner, and thrust sometimes for miles over another series. Rocks that have been profoundly altered in this way are called *metamorphic* rocks, and such rocks bulk largely in the Scottish Highlands.

The whole succession of the sedimentary rocks is

divided into various classes and sub-classes. Resting on the very oldest rocks there is a great group called Primary or Palaeozoic. Next comes the group called Secondary or Mesozoic, then the Tertiary or Cainozoic, and finally a comparatively insignificant group of recent or Post-Tertiary deposits. The Palaeozoic rocks are divided again into systems, and since the rocks of Renfrewshire fall entirely under this head, we give below the names of the different systems, the youngest on top.

Palaeozoic Rocks.

Permian System.
Carboniferous System.
Old Red Sandstone System.
Silurian System.
Ordovician System.
Cambrian System.

The old rocks, particularly the Ordovician and the Silurian, are typical of the Southern Uplands, but we have already seen that Renfrew lies entirely in the Central Lowlands, and therefore these systems are not represented in the county. With the exception of a limited area where the Old Red Sandstone rocks are found, the rocks of Renfrew belong entirely to the Carboniferous system. The Old Red Sandstone rocks are confined to the south-west of the county, and can be particularly well seen along the shore from Wemyss Bay to Inverkip. They stretch inland on both sides of the county boundary and point a finger north-eastwards that almost reaches Loch

Thom. For the most part they are sandstones of a bright red colour, and have been quarried largely for building-stone. Almost all the houses in the locality have been built from the local Old Red Sandstone series.

The Carboniferous system is by far the most important in the county, for it occupies nine-tenths of the total area. Its sub-divisions are as follows:

> Coal Measures.
> Millstone Grit.
> Carboniferous Limestone Series.
> Calciferous Sandstone Series.

The true Coal Measures and the Millstone Grit[1] are not found in the shire, the geology of which is therefore seen to possess no great variety, although abounding in interesting and difficult problems. The Carboniferous Limestone series consists of sandstones, shales, limestones, coal-seams, and ironstones, and forms most of the lowlands in the east of the county. These rocks are not the true Coal Measures but they contain coal-seams of considerable value. All the coal worked in Renfrewshire comes from this series. The sandstones of the Carboniferous Limestone group furnish building-stone of the finest quality, and a large proportion of the city of Glasgow has been built from the sandstones found in the neighbourhood of Giffnock. The series falls into three

[1] The Geological Survey Maps show patches of Millstone Grit near Barrhead, but the lower boundary of that formation is now taken at the Castlecary Limestone instead of the Arden Limestone, and thus the rocks formerly classed as Millstone Grit become the top members of the Carboniferous Limestone series.

sub-divisions : (*a*) Upper Limestone Group, containing limestones and thick sandstones ; (*b*) Middle Group, containing several workable seams of coal and ironstone associated with sandstones and shales, but not with limestones; (*c*) Lower Limestone Group, containing limestones and sandstones. The Giffnock Sandstones form a part of the Upper Limestone series which stretches north-east from the volcanic rocks near Barrhead, past Giffnock to Strathbungo, where a great fault occurs, on the other side of which the rocks have sunk till the Coal Measures have been brought level with the Carboniferous Limestones. Sandstone is the chief mineral obtained from these rocks, but here and there a coal-pit occurs working the coals of the Middle Group, and at one or two places, Darnley particularly, the limestones of the Upper Group are quarried.

The Calciferous Sandstone rocks consist largely of sandstones and shales, but the most remarkable feature of this period was the wonderful outbreak of volcanic activity that took place over most of the lowlands of Scotland. Hundreds of volcanoes burst into action, hurling forth stones and ash, while molten lava gushed from their craters, until in this way thousands of feet of rock were built up. From Stirling to Campbeltown these rocks are found, forming the Kilsyth Hills and the Campsie Fells, the Kilpatrick Hills, the Kilbarchan Hills, the Corkindale Moors, and the Eaglesham Moors, stretching thence along the borders of Lanarkshire and Ayrshire, and forming the high ground between these counties. The volcanic series is perhaps best developed in Renfrewshire,

and in certain parts of the county, notably in the hills
round Misty Law, many of the vents from which the
molten rock poured forth have been recognised.

These volcanic rocks are hard and resistant to the
weather, and it is simply for this reason that they now
form the hills of the county. They were poured out in
great horizontal sheets, and this explains the plateau
nature of the hills formed by these rocks. The series
was not built up in a single outbreak. Eruption succeeded eruption, each pouring forth a new flow of lava.
Thus the rocks were built up in a series of layers, and the
resulting step-like outlines of the hills form one of the

Diagram showing effect on hill outlines of volcanic
lavas overlying softer sedimentary rocks

most characteristic features of the scenery of the Clyde
area. Looking from Ashton towards the ridge that
terminates at the Cloch Point we see this well exemplified.
On the left the junction between the lavas and the underlying sandstones is quite evident from the sudden steepening
of slope that takes place there. On the sky-line the
volcanic rocks are seen to descend in a series of steps, each
step marking a different lava flow. In fact, wherever
tabular hills assume this peculiar stepped profile, one may
guess with some confidence that the hills are of volcanic
origin.

Here and there throughout Scotland there occur isolated hills in the form of steep crags, of which the best known are Edinburgh Rock, Stirling Rock, and Dumbarton Rock. These hills are geographically and historically of the utmost importance, for, since the earliest times, they have formed strongholds on which castles were built, and round which in time busy towns grew up. Such a town—Dumbarton is an excellent example— owes its existence in every case to the fact that many ages ago, long before man appeared on the earth, a volcano happened to break forth and filled its chimney with solidified lava or hard, fragmentary rock. The surrounding softer rock was eroded more rapidly than the resistant volcanic material, and the denuded "neck" or "stump" of the old volcano remains as an isolated crag. Several examples of this type of hills occur in Renfrewshire. One of the best, although not on a very large scale, is the picturesque Craig of Carnock, near Barrhead. Neilston Pad, the hill that presents such a curious, artificial appearance, even when seen from as far off as Glasgow, has had a similar origin. The rock is of quite a different type from the surrounding lavas, and breaks through them in the same way as the more common neck pierces the surrounding sedimentary rocks. The rock composing Neilston Pad has been discovered to be a trachyte, a species hitherto entirely unknown in the west of Scotland[1].

[1] The discovery is due to Mr Tyrrell of Glasgow University. The results are not yet published, but the writer has been over the ground with Mr Tyrrell and has had the privilege of examining his micro-sections.

Volcanic activity in Renfrewshire manifests itself in still another way. At different periods in the earth's history, the solid crust has cracked along fissures, sometimes many miles in length, and up the crack has welled the molten lava, finally solidifying in a wall-like form called a dyke. These volcanic dykes are frequently harder

Volcanic Dyke at Ashton

than the surrounding rocks, and therefore stand up like artificial walls in a manner that eminently justifies the name. They are very plainly marked in the sandstones about Wemyss Bay, where the black bands of volcanic rock contrast strongly with the bright red colour of the surrounding sandstones. One of the largest can be seen on the shore a mile south of Ashton where it forms a vertical wall over a dozen feet in height.

After the deposition of the Carboniferous rocks the geological history of Renfrewshire for many ages is a blank. If we compare the story of the rocks to a book of history, we may say that many of the later chapters of the work have been torn out and completely lost. Certainly many different systems were laid down on the Carboniferous rocks ; doubtless the area was at times dry land, at times covered by the deep waters of the sea, but all the succeeding strata have been stripped away by those two all-powerful co-operators in destruction—time and the weather. The last chapter of the record tells us of the ice-age. For a long time the climate had been growing more severe. Tropical plants and animals were supplanted by temperate, and then by arctic, forms, and finally a great ice-sheet occupied all the higher parts of Scotland. Huge glaciers crept slowly down the valleys from their gathering grounds, the extensive ice-fields of the Highlands and the Southern Uplands. The glaciers have gone, but their work remains to tell their story—the grooves and scratches on the rocks, the excavation of lake-basins and the deepening of valleys, the moraines, well nigh as perfect now as when they were thrown down. The scratches on the volcanic hills make it plain that most of Renfrewshire was over-ridden by an ice-sheet that came from the north-west. Professor Gregory, however, has recently shown that in the Loch Libo Gap, during one period at any rate, there was a movement from south-west to north-east, and suggests that the glacier moving through the pass came

from the Hill of Stake region and was deflected to the
north-east, or possibly came even from Arran. Sir
Archibald Geikie, too, has recorded that pieces of Old
Red Sandstone and other rocks found far to the south are
occasionally met with in Renfrewshire, and explains this
by an alternate advance of the ice from the Highlands
southward, and from the Southern Uplands northward.
The thick sheets of tough clay studded with boulders,
found all over the county, are attributed to the ground-
moraine of the great ice-sheet. During the melting of
the ice, accumulations of sands and gravels were occasionally deposited. A fine series of such mounds is to be seen
near Eaglesham.

At the close of the ice-age an event occurred of
fundamental importance to the future welfare of the
towns on the coastal fringe of Renfrewshire. This was a
rise of the land or (we are not sure which) a withdrawal
of the waters of the sea. This converted the old sea-
beach into dry land, and thus formed a narrow band of
low, flat ground round the coast, an eminently suitable
site for watering-places. All the coast towns of the Firth
of Clyde are situated on the "raised beach," and at most
of them can be seen the old sea-cliff against which the
waves once dashed, now left high and dry a few score
yards inland. In Renfrewshire this can be well seen
along the road between Gourock and Ashton. The road
itself and all the houses next the sea are situated on the
raised beach, which is flanked inland by low cliffs obviously
of marine origin.

The volcanic rocks of the Renfrewshire hills abound

in rare minerals. Most of these are of interest chiefly to the mineralogist, but one or two are worth particular notice. Copper is found in more than one locality. In a whin-stone quarry near Barrhead, films and strings of metallic copper sometimes come to light, while the sandstones behind Gourock are impregnated with copper ore. The metal has even been mined in the volcanic rocks near Lochwinnoch. Fluor-spar, an exceedingly beautiful and somewhat uncommon mineral, is found in the county. It is familiar to many people as the "blue John" of Derbyshire, where the workers maintain with pardonable inexactitude that their mine is the only known locality for the mineral. Beautiful little crystals of a pale green or blue colour may still be obtained from the quarry in the igneous rocks behind Gourock. When the railway tunnel through the volcanic rocks at Bishopton was being made, minute crystals of an exceedingly rare mineral were discovered. The crystals are yellow and lustrous, and the largest are tiny pyramids a quarter of an inch high. The mineral received the name Greenockite after Lord Greenock, and some of the finest specimens are to be seen in the Hunterian Museum, Glasgow University.

The soil of the low parts of the shire is generally a rich alluvium which is capable of bearing heavy crops. Along the larger valleys also a fertile alluvial loam is found, although in places the lack of adequate drainage sours the soil, which is thus used only for grazing purposes. Over the volcanic rocks the soil is always very thin, the bare rock frequently protruding, except where sheets of boulder-clay occur. The thinness and the height above sea-level of this soil prevent its use in agriculture, but in itself it is of considerable fertility. The course of a volcanic dyke, for example, can often be traced by the line of flourishing trees it carries. In many places the volcanic hills nourish a thick, springy turf, which when well cut forms the nearest approach to sea-side turf that can be found in any inland district. Thus golf courses have been instituted on the hill-sides (the Fereneze Club, the Cathkin Club, the Milngavie Club), the turf of which is much superior to that of the ordinary inland green, owing its excellence to the nature of the underlying rocks. In view of the enormous increase in the number of golfers in recent years, there is a hint here worthy the notice of the organisers of new courses.

7. Natural History.

Many centuries ago the British Isles formed a part of the continent of Europe. Where the waters of the English Channel now ebb and flow, there was dry land offering a free passage to the migration of plants and animals from Central Europe to this country. Such was the case when the palaeolithic hunters, the men who chased the mammoth and the reindeer with their rude stone weapons, lived in Britain. By neolithic times, however, when our primitive ancestors were using finely chipped and polished weapons of stone, the British Isles had become separated from the Continent, and Ireland was severed from Great Britain. The land-bridge existed after the disappearance of the

great ice-sheet from this country, and plants and animals
from Europe migrated to Britain. The land connection,
however, did not remain long enough for all the continental forms of life to make their way to Britain, for we
find that there are fewer species in Great Britain than in
Western Europe, and fewer species in Ireland than in
Great Britain. For example, every one knows that there
are no snakes in Ireland, and for this, Saint Patrick is
generally given the credit ; but an alternative hypothesis
is that their absence is due to the breaking of the land-
bridge before migration to Ireland could take place. The
comparative poverty of animal species in Britain is most
marked in the case of the mammals and the reptiles, since
these do not possess the power of flight. Thus while
Germany has about 90 species of land mammals, Britain
has only about 40. There is not a single species of mammals, reptiles, or amphibians found in Britain that is not
found on the Continent; and only one bird, the common
red grouse of Scotland, does not occur in continental
Europe.

The mammals of Renfrewshire are typical of Scotland
as a whole. The common bat and the long-eared bat are
frequently met with. Daubenton's bat is rare but has
been found at Craigenfeoch, while the rare lesser horse-
shoe bat has been seen at Crookston. The hedgehog and
the mole are abundant, the common shrew does not belie
its name, but the water-shrew is rare. In most of the
lowland counties of Scotland the wild-cat is now extinct,
but a specimen was noted at Gleniffer in 1895. The
pole-cat, however, has quite disappeared, the last recorded

occurrence being at Craigenfeoch about 1868. The badger and the pine-marten have vanished, but the fox, the stoat, the weasel, and, in suitable places, the otter, are still common. Most of the British rodents occur in Renfrew. As in the neighbouring counties, the harvest mouse is quite unknown, but strange to say, a nest was found at Kilbarchan in 1895. The old black rat is extinct. He has been everywhere pushed out of existence by his interloping relative, the brown rat. Mice and voles are everywhere common; and in many of the woods the brown, bushy tail of the squirrel may frequently be seen vanishing round a tree trunk. Rabbits and hares are ubiquitous. Three hundred mountain hares were killed during one season on the Misty Law Hills.

Many parts of Renfrew offer favourable opportunities for the study of bird life. It would be tedious to enumerate all the species of birds common in the county. It is enough to note that the diversity of land surface and the frequent occurrence of sheets of water are reflected in the variety and abundance of the bird life. On the high moors the mournful cry of the curlew or the lapwing can everywhere be heard, but the number of other species is comparatively small. Near the well-wooded banks of the Cart, however, the number not only of species but of individuals is remarkable. Along the coast south of Gourock sea-birds abound, although they do not nest so frequently as they do farther down the firth. The numerous lochs and dams of the county, again, are the favourite haunts of duck and teal, coot and moorhen, grebe and gull.

Compared with the Continent the reptiles and the amphibians of Britain are remarkably few in number. Most of the British species are found in Renfrewshire. There are two species of lizards, the lizard proper (*Lacerta vivipara*), and the blind-worm or slow-worm. The former may often be seen on a hot day frequenting dry, sunny places such as stone-heaps, walls, or ruined buildings. The latter is not so often met with, but may sometimes be seen among dead wood, decayed leaves, or stone-heaps, generally preferring a dry situation. The blind-worm is of course not a snake, as is often supposed. It is an inoffensive, timid, and perfectly harmless creature. When caught it becomes so rigid with fear that it easily breaks in two. It is from this fact that its specific name *fragilis* is derived. Of the true snakes the adder or viper is our only common representative. It is the only poisonous reptile in the country. To the healthy adult its bite is practically never fatal, although deaths have resulted in the case of children and infirm persons. The adder loves dry, warm places, among ruins, or under fallen trees, or on sunny banks. The frog, the common toad, and the newt are everywhere abundant.

The plants of Renfrewshire are fairly representative of the whole of Scotland. There is, however, no mountain of sufficient height to exhibit well the peculiar alpine flora of Scotland, found on Ben Lomond and other mountains in the lower basin of the Clyde. The old Caledonian forest probably existed over many areas that are now bare of trees. The existing woods of Renfrewshire have practically all been planted by man. Some of the finest specimens of trees in all the west of Scotland are to be found in the grounds of Blythswood, Erskine, and Pollok. The uplands of the county are mainly moor and marsh. In autumn they are purple with the flowers of the ling and the heath. The milk-wort, the bog asphodel, and in wetter parts, the cotton grass, are abundant. In the marshes also the butter-wort and the sundew set their traps for unwary insects. All summer the grassy uplands are bright with the tiny, yellow flowers of the tormentil, and the gaily-coloured mountain-pansy. The sunnier slopes of the Gleniffer and Fereneze Braes are especially beautiful with these exquisite little flowers. The hedge-rows of Renfrewshire are not nearly so rich as those of the border counties where the climate is more genial. In addition the low-lying parts of the shire are generally covered with boulder-clay, which gives a stiff, cold soil that is very unfavourable to variety of plant life.

8. The Coast Line.

It would not be easy to say at what point in Ren-frewshire the bank of the river ends and the coast of the estuary begins. For navigation purposes the river merges into the sea at Greenock, but a distinct widening of the waters occurs about Langbank, which may therefore be taken as the starting-point of a tour round the Renfrew-shire coast. If the tide is in, covering the ugly mud banks in the foreground, the picture across the water is altogether charming. Here is the best place to view Dumbarton Rock,

still grim and threatening as when the tide of battle ebbed
and flowed around it (see p. 109). The low Vale of Leven
is flanked by volcanic hills, while in the background
Ben Lomond stands head and massive shoulders above his
Highland neighbours. We walk along the old raised beach
with low cliffs on our left, against which in ages past beat the
waves of the sea. Now we look across to Cardross, where
the Bruce spent the evening of his days. His ships must
often have cleft the waters at our side. Inshore are the
floating wood-yards that indicate our nearness to Port
Glasgow. On the outskirts of the town is Newark Castle,
the old baronial dwelling of the Maxwells. We hurry
through the dingy main street of Port Glasgow, our ears
assailed by the incessant, clattering fusillade from the ship-
yards, and make for Greenock. The navigable channel
of the estuary, guarded by lights, keeps close to the Renfrewshire coast, and a constant stream of shipping marks
the course.

Our approach to Greenock is fittingly heralded by its
finest harbour, the great James Watt Dock, the culmina-
tion of a magnificent series of enterprises in harbour construction. More perhaps than any other town in Scotland,
Greenock owes its prosperity to one family, the Shaws,
lairds of the town. The eastern suburb of the town is
Cartsdyke, famous in former times for its red-herrings.
It was here in 1696 that a ship was fitted out in order to
take part in the ill-omened Darien Scheme. Nowadays
the native place of James Watt seems entirely devoted
to engineering and industrial enterprises, yet its literary
associations are neither few nor uninteresting. John Galt

Highland Mary Monument, Greenock

died here, and here in the old kirk-yard lies Highland
Mary. To Greenock as schoolmaster came John Wilson,
the author of *The Clyde*, his philistine employers sternly
banning the profane art of poetry. At the west end of
the town is Princes Pier, one of the main gateways of the
summer tourist traffic. In the broad estuary opposite the
pier a submarine shoal stretches out a long arm that is
known as the Tail of the Bank. Fort Matilda stands at
the end of the beautiful esplanade, and looks across the
firth to Kilcreggan less than two miles away. Should
a hostile man-of-war by some chance escape the forts
lower down the firth, and keep close to the Kilcreggan
shore to avoid the guns of Fort Matilda, it would find
itself between Scylla and Charybdis, for those innocent,
smooth mounds along the shore mask the black muzzles
of another fortress.

We round the bend into the beautiful Bay of Gourock.
The sheltered blue waters are dotted over with pleasure
craft of every kind. The sun shines white on the sails
of the dainty little two and a half raters, and sparkles on
the gleaming brass-work of palatial steam-yachts. The
creamy-waked river-steamers dash up and graze the piers
with an *abandon* that would make an English river-
captain gasp in horror. It is an intrusion of volcanic
rock that has caused the bay. It is sheltered from the
western gales by Kempoch Point, which juts into the sea
forming a natural break-water, because the igneous nature
of the rock makes it harder than the surrounding sandstones. Granny Kempoch, a monolith of mica-schist six
feet high, stands between the cliff and the castle. For ages

Yachting at Gourock

it has been an object of superstitious reverence. Seamen
and fishermen would pace seven times round it singing in
order to ensure a prosperous breeze. From "the Goraik"
James IV put to sea in 1494 on his expedition to the
Western Isles to subdue the Hebridean chiefs. The

Old Granny Kempoch, Gourock

shore-road runs south-west, past the pleasant villas of
Ashton, with cliffs and sea-caves on the left, and seaward
a glorious outlook up Loch Long and the Holy Loch.
A few yards off the high road is the ancient ruined Leven
Castle, a former stronghold of the Mortons and the

4—2

Sempills. By the side of the castle a little road runs up the hill from the shore. It is easy to find, for it is lined with minatory boards invoking many and grievous penalties on the head of the unlucky trespasser. Walk up this road, pass through the farm and up the hill side as far as the track continues, and, on looking round, there shall lie before you one of the finest panoramas in all Scotland. Some of the peaks that are visible are shown on p. 17, but one other interesting feature may here be mentioned. The even crest-line of the low hills to the north is deeply notched just behind Kilcreggan. This marks one of the most important geographical features in Scotland. It is the line of the great fault that separates the Highlands from the Lowlands.

A mile past Leven Castle we reach the white tower of the Cloch Lighthouse. On clear nights the steady white beam of the Cloch stabs the darkness for some 15 miles, while in times of fog, the mournful wail of its steam-horns sends eerie messages to the dwellers on the opposite Cowal shore. Cloch Point, like Kempoch, is produced by the occurrence of hard rock among the softer sandstones. The bedded lavas of Renfrewshire reach the coast only at this point, and this makes it salient. The broad sandy stretch of Lunderston Bay brings us to Inverkip. In the seventeenth century the county was sorely plagued by witches, as we shall see later, and this district was the scene of some of their pranks. South of Inverkip the road swings inland skirting the grounds of Castle Wemyss, the home of Lord Inverclyde, head of the famous Cunard line of steamships. A little farther and we pass the road to Kelly House where lived James Young, founder of the paraffin-oil industry of Scotland. It is but a few steps now to Wemyss Bay pier and the Kelly Burn, south of which we enter Ayrshire. From Inverkip to the boundary we have walked on Old Red Sandstone rocks, the bright red

Castle Wemyss

of which is crossed every now and then by a black band marking the course of an igneous dyke. Just on the county boundary the railway stops, and the passengers pour on to the decks of the speedy paddle-boats that wait to take them to their longed-for summer havens.

9.　Weather and Climate.

The weather of Britain depends largely on the distribution of atmospheric pressure over these islands. To put
the matter in its simplest form, when the barometer is
high we expect good weather, and when the barometer
is low we expect wet and stormy weather. These two
types of weather correspond respectively to a condition
of high atmospheric pressure or anticyclone, and a state
of low atmospheric pressure or cyclone. The winds in
a cyclone are often strong, and swirl round the centre of
lowest pressure in great spirals with a direction opposite
to that of the hands of a clock. When anticyclonic
conditions prevail, the winds are light and move round
the area of highest pressure in the same direction as the
hands of a clock.

Generally speaking, we may say that the winds of
Scotland throughout the year are controlled by three
fairly permanent pressure centres. There is a low pressure area south of Iceland, an Atlantic high pressure area
about the Azores, and a Continental area in eastern Europe
and west Asia, that is high in winter time and low in
summer time. In winter as a rule the Icelandic and the
Continental centres predominate, as they are then working
in harmony. The tendency of both centres is to draw the
air in a great swirl between them from south-west to north-
east. Therefore we find that in winter south-west winds
predominate in Scotland.

Occasionally the Continental anticyclone spreads as
far as Scotland, and then for a few days in winter we

experience clear skies (with dense fog in towns), keen frosts, and very light winds. All too soon the Icelandic cyclone centre reasserts itself, and we are back again to storms of sleet or rain with a higher temperature. In summer the Atlantic high pressure centre has more influence. It tends to draw the winds more to the west, sometimes to north-west. This high pressure area with its accompanying fine weather is now at its most northerly limit, and occasionally spreads over these islands, reaching the south of England frequently, but not so often extending to Scotland.

To sum up then, we find that on the whole the prevailing winds of Renfrewshire are westerly and south-westerly. In winter, south-west winds are by far the most common, and our heavy gales are nearly always from the south or south-west. This is due to the presence of a cyclone or area of very low pressure to the north-west of Britain. The wind whirls round the low pressure centre and thus a south-west gale is experienced in Britain. In summer there is a shift of the winds towards the north, with the result that winds from the west predominate. Easterly winds are commonest in late spring and early summer. In May they are more frequent as a rule than winds from any other direction. In many parts of the country the trees are inarticulate but convincing recorders of the prevailing wind direction. They grow with their branches pointing east or north-east away from the wind. The branches of the tree shown in the photograph on p. 56 point almost exactly north-east.

It is a general belief in this country that storms are more frequent and violent at the time of the equinoxes than at any other time. The phrase "equinoctial gales" is heard so frequently that the assumption it implies is accepted without question. It is an interesting point, therefore, to consider if the phrase is truthful. Examination of actual records proves that the so-called equinoctial gales are mythical. Storms are not more frequent at the equinoxes than at any other time. This has been clearly

Tree near Barrhead showing S.W. wind

shown in America, where the myth is also well established; but the splendid series of weather records kept at Glasgow University Observatory during the last forty years are quite convincing on the point. They show that storms are most frequent in winter and least frequent in summer. The maximum number occurs in January, and the number decreases steadily till June and July, then rises steadily again to January.

The prevailing south-west winds of this country in winter have much to do with our favourable winter climate. The climate of the British Isles in winter is milder than that of any other part of the world in the same latitude. The following comparison will illustrate this very strikingly. Aberdeen and Nain (Labrador) are in the same latitude. The mean temperature of the coldest month at Aberdeen is 35° F., or *three degrees above the freezing-point*. The mean temperature of the coldest month at Nain is −4° F., that is *thirty-six degrees below freezing-point*. Most of us learned at school that our good fortune as regards climate was due to the beneficent influence of the Gulf Stream, but in recent years this explanation has been abandoned. It is a myth as fanciful as the supposed equinoctial gales. The Gulf Stream becomes a negligible factor a little to the east of the Newfoundland Banks. Our true benefactor is the wind. In winter time the south-west winds blow from the warm southern regions of the Atlantic, raising the temperature of Britain, and depositing moisture, which means a still further rise owing to the liberation of the latent heat. In addition, they blow the warm surface waters of the ocean from more southerly latitudes, and cause them to flow round and past our islands. There is no strongly marked current, but a general "Atlantic Drift" of the heated surface waters.

The temperature conditions of Renfrewshire are similar to those of other counties on the western slope of Scotland. The summers are cooler, and the winters are milder than on the east coast. The mean temperature for Paisley in January, taking an average over 24 years, is

39° F., and the mean temperature for July is 59° F., giving a mean annual range of 20° F. The mean annual range for Edinburgh is 21° F., and for London is 26° F.

Renfrewshire is not so favoured in the way of sunshine as many other parts of the country. The amount of sunshine diminishes as we go from south to north or from east to west. The average number of hours of sunshine per annum at Paisley is 1201, while on Ben Nevis the amount is less than two-thirds of this figure, namely 735 hours. Aberdeen on the other hand has 1401 hours of sunshine per annum. The temperature and the sunshine are important factors in crop-raising. For example, wheat needs a hot bright summer to ripen properly, and therefore we find that Renfrew is not an important wheat-growing county. Fife is twice the size of Renfrew, but it grows more than six times the amount of wheat that Renfrew does.

The records of rainfall for Renfrewshire are neither so numerous nor so trustworthy as could be wished. Practically all the older records are unreliable. For example, the writer of the section on Renfrewshire in the *New Statistical Account* of 1845 gives the average annual rainfall of Greenock as 35 inches. More recent observations, however, taken over a period of 25 years, show an average rainfall of nearly double that amount, namely 65 inches. Either the rainfall of the county has altered to an amazing extent or the early records are untrustworthy, and the latter is the likelier explanation. In the same way it is stated that there are no fewer than three climates in the single parish of Neilston. One of them "begins at the

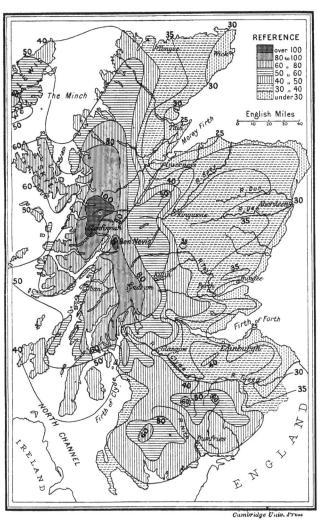

Rainfall map of Scotland
(*By Andrew Watt, M.A.*)

parting of the roads to Neilston and Irvine. No one ever
came to the separation of the two roads above mentioned
who did not feel immediately a sensible difference, let the
weather be what it may." One must suppose that the

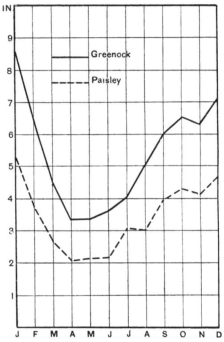

Rainfall throughout the year at Greenock and Paisley

reverend gentleman's meteorological sense (or his imagi-
nation) was preternaturally acute, or that the abrupt
climatic change he mentions has migrated to some other
corner of the road.

In Renfrewshire as a rule the driest month of the year is April, and the wettest is January. This can be clearly seen from the curves on p. 60, which show how the rainfall varies from month to month. The curves show the average rainfall for each month of the year at Paisley and Greenock. Although the total amounts for the year are quite different, yet the fluctuations from month to month show a similarity that is astonishing. Spring is much the driest season of the year, and winter is the wettest. The curves show very clearly the marked rise in the rainfall that takes place in July and August, a phenomenon that is but too well known to holiday-makers in the west. As regards length of daylight, dryness, and hours of bright sunshine, June is undoubtedly our ideal month of summer.

The following table, compiled from the annual volumes of *British Rainfall*, shows the average rainfall over the ten years 1900–1909 of several selected stations in Renfrewshire:

Station	Height above sea-level	Rainfall
Crookston	about 50 feet	37.9 inches
Paisley Observatory	107 ,,	44.1 ,,
Bishopton	195 ,,	46.6 ,,
Lochwinnoch	240 ,,	47.4 ,,
Waulk Glen	280 ,,	49.3 ,,
Gryfe Reservoir	640 ,,	66.3 ,,
Loch Thom	643 ,,	67.5 ,,

The foregoing table illustrates beautifully the effect of altitude on rainfall. The stations are arranged in

order of height above sea-level, and it is seen that without exception the rainfall increases with increasing altitude.
If the averages in the above table had been taken over a
much longer period it would probably have been found
that in each case the rainfall was an inch or two less.

10. The People — Race, Language, Population.

The earliest inhabitants of Britain probably crossed
from the continent of Europe when it was connected to
these islands by a land-bridge. They used very roughly
made stone weapons and were mighty hunters, chasing
the reindeer, the mammoth, the wild horse, and other
animals that lived in this country in those days. From
their stone weapons they are called palaeolithic (ancient
stone), and their nearest representatives in modern times
are believed to be the Bushmen of Africa. Authorities
are almost unanimous in maintaining that there is no
evidence that this race reached Scotland. These early
palaeolithic men were followed by a race that used stone
weapons of a much finer type (neolithic), and relics of this
race are found all over Scotland.

One of the most constant and valuable physical char-
acters of a race is the shape of the skull, which may be
classed as long or broad. The primitive race of Scotland
was long-skulled, short in stature, and probably very dark
in complexion. They are known as Iberians, and have
few affinities with Celts or Teutons. Later on Scotland
was invaded by Celtic tribes, who were broad-skulled, and

who are generally supposed to have driven out or exterminated the Iberian race, for in early historic times the language of almost the whole of Scotland was Celtic, with, however, a number of non-Aryan peculiarities of syntax. Yet it is a remarkable fact that the majority of the people in Scotland at the present time are long-skulled, and therefore cannot possess much Celtic blood. Now the Teutons are long-skulled, but we know from history that the Scottish Highlanders are not of Teutonic stock, and in addition the Teutons are fair, while the Celtic speaking races are very much darker in complexion than the people of other districts. It would seem therefore that the Celtic invaders were merely a predominating and ruling caste, who imposed their language and culture on the conquered tribes, but did not seriously dilute their blood. The aboriginal stock absorbed the invaders, and thus on the whole the inhabitants of Scotland may be said to be of pre-Celtic and Teutonic blood. No definite agreement on these points, however, has yet been reached.

The earliest records relating to the Clyde valley state that it was in possession of the Damnonii, a Celtic-speaking tribe. At the end of the fourth century, when the Roman legions were withdrawn, Clydesdale was inhabited by the Scots, a Goidelic tribe, and the original inhabitants were driven to the south of the district. About the beginning of the fifth century the Teutonic race began to appear in Scotland, and for five hundred years this immigration went on, until practically the whole of the Lowlands was in the hands of Teutonic tribes, the ancestors of the present Lowland Scots.

The place-names of Renfrewshire are not so purely Gaelic in origin as those of many other parts of Scotland. The names of the hills illustrate this. The Celtic *bens*, *stobs*, *sgors*, *maols*, and *mealls* are as a rule conspicuous by their absence, although *dun*, *knock*, and *cairn* are found. The Anglo-Saxon words *hill* and *law* are much more common, while *rig* and *dod* (hopelessly unmusical compared with the sonorous Gaelic) are more sparingly met with. Celtic words of course occur all over the county, as is only to be expected from the fact that Renfrewshire formed part of the old Celtic kingdom of Strathclyde. *Kil*—a church, *auch*—a field, and *inch*—an island, are perhaps the commonest Gaelic words found in combination. Of the town names it is natural to find that the older are of Celtic origin while the newer are Anglo-Saxon.

We have seen in a former chapter that Renfrewshire ranks twenty-seventh among the counties of Scotland in size, but in population there are only two above it. Putting this in another fashion we may say that while the county comprises $\frac{1}{125}$ of the total area, it contains almost $\frac{1}{15}$ of the population of all Scotland. The actual figures from the census of 1911 are 314,574 for Renfrew out of a total population of 4,759,445. It follows that the county must be densely populated in spite of the large areas of comparatively useless hill country. In this respect it takes third place also, ranking after Lanark and Edinburgh. There are 1290 persons to the square mile, and only Renfrew and the other two counties mentioned run into four figures for population density. In this respect it forms a striking contrast to Sutherland which has but 9½ persons to the square mile, and even to all Scotland which has 157. It is only since the "Industrial Revolution" that the population of Renfrewshire has increased so much. Before the application of machinery to

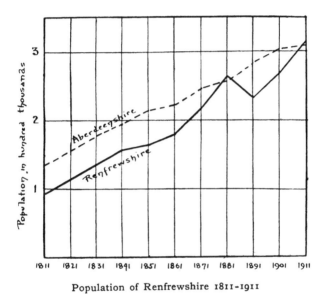

Population of Renfrewshire 1811-1911

(*The sudden fall between* 1881 *and* 1891 *was due to a change in the county boundaries*)

industries in the second half of the eighteenth century, the county did not contain a tenth of the inhabitants it now possesses. This is strikingly shown by the fact that a hundred and fifty years ago the population of the whole county was only one-third of the population of Paisley at the present time.

The alien element is by no means strong in Renfrew. Compared with Lanark, for example, which contains more than half the foreigners in Scotland, the numbers are insignificant, reaching a total of 646 in 1901. This gives a percentage of about a quarter, compared with one-half per cent. of aliens for all Scotland. Italians and Germans preponderate, forming nearly two-thirds of the total number, while in 1901 there was but one Asiatic, and not a single representative of Africa. But there are others, not foreigners, whose knowledge of English is insignificant or absent. At last census there were 41 persons who spoke Gaelic only, forming literally a Celtic fringe, for 25 of them lived in Greenock.

The occupations of the people are numerous and varied. Naturally industrial pursuits claim by far the majority of the workers. Considering males alone, such workers form 73 per cent. of all the people occupied, commerce ranks next with 17 per cent., agriculture and fishing claim 4 per cent., while professional men are a trifle fewer. The number of miners in Renfrew is remarkably small. There are only between seven and eight hundred compared with over fifty thousand in Lanark. Naturally the conditions are different with women. Household duties, no matter how heavy, receive no salary, and are therefore not considered "work" by the census, so that nearly seventy thousand women are (nominally) unoccupied. In Paisley, as might have been expected, the proportion of female to male workers is abnormal. Between a third and a half of the industrial workers of all Scotland are female, but in the textile industry of Paisley the proportions are exactly reversed, the females outnumbering the males by nearly three to one.

11. Agriculture.

Although at the present time Scottish gardeners and Scottish farmers have a world-wide reputation, yet it was not till the eighteenth century that there was any agriculture worthy of the name in Scotland. Most of the country was unenclosed, roads and bridges were almost unknown, artificial drainage was not employed, and only the driest parts were tilled. Yokes of oxen dragged a rude plough far up the hill-sides, because the lower parts were hopeless swamps. A few sentences may be quoted from Henry Grey Graham's description of the state of agriculture at the beginning of the eighteenth century. "There were no enclosures, neither dyke nor hedge between fields, or even between farms; so that when harvest began or the cereals were young, the cattle either required to be tethered, or the whole cattle of the various tenants were tended by herds."…"When the harvest was over the cattle wandered over all the place, till the land became a dirty, dreary common, the whole ground being saturated with the water which stood in the holes made by their hoofs. The horses and oxen being fed in winter on straw or boiled chaff, were so weak and emaciated that when yoked to the plough in spring they helplessly fell into bogs and furrows, even although to fit them more thoroughly for their work they had been first copiously bled by a 'skilful hand!'"..."The harrows, made entirely of wood,—'more fit,' as Lord Kames said, 'to raise laughter than to raise soil,'—had been in some districts dragged by the tails of the horses, until the barbarous practice was condemned by the privy council."..."If one man dared to cultivate a neglected bit of ground, the others denounced him for infringing on their right of grazing on the outfields. How could he begin the growing of any new crop? The others viewing every innovation with the contempt which comes from that feeling of superiority, which ignorance and stupidity produce, would refuse to join him."..."With a system so atrocious, with land uncleaned, unlimed, unmanured, undrained, it frequently happened that the yield could not feed the inhabitants of the district, and men renting from 40 to 100 acres needed to buy meal for their families."

Gradually new crops and better methods were introduced. The cultivation of turnips and potatoes marked the beginning of a more rational agriculture. Old ideas, such as the determination to use no mechanical aids to winnowing, because it contravened the Scriptures, and "was making Devil's wind," gradually disappeared. Stockbreeding was introduced, the land was let in larger holdings, alternation of crops was practised, artificial fertilisers were used, until at the beginning of the nineteenth century agriculture was on a satisfactory basis.

In some respects Scotland will always be at a disadvantage compared with England. In many parts the soil is as fertile as any south of the border, but the more favourable climate of England causes an earlier harvest. An additional crop of turnips or cabbages or vetches can then often be secured after the main crop has been got in, whereas in Scotland this can very seldom be done.

Renfrewshire compares favourably with Scotland as a whole with regard to the proportion of its surface under cultivation. About three-quarters of Scotland is uncultivated, whereas the proportion for Renfrewshire is only about two-fifths. If we make allowance for its size, the county is, like Wigtown, one of the greatest dairy-farming shires in Scotland and possesses about 165 cattle to every thousand acres. The causes are mainly geographical. The western side of Scotland is rainy, and rich grass is therefore easily grown, while the decomposition of the volcanic rocks of Renfrew produces a loam that nourishes a thick, sweet pasture. The cattle are kept chiefly for dairy purposes, and therefore the great majority of them are Ayrshires. This breed has been found peculiarly suitable to the moist climate of the south-western counties. It is not only hardy, but yields a larger proportion of milk to food consumed than any other breed in the country. Glasgow and other large towns absorb the supply of most of the dairy-farms, and cheese-making is consequently not so general as it once was, except for home use. In the number of horses, too, Renfrew compares favourably with its neighbours Ayr and Lanark. In every thousand acres in Renfrew there are 21 horses, while Lanark has 16, and Ayr 13, per thousand acres.

On the other hand sheep-rearing is not of special

importance in the county. There are some 280 sheep to each thousand acres, whereas in Roxburgh there are in the same area more than four times as many. The stock consists chiefly of Cheviots and black-faced sheep. The wool of the black-face does not bring so high a price as that of the Cheviot, but the former breed is hardier and more suited to hilly tracts. It will thrive on poor fare and withstand privations that would exterminate any other breed. In 1910 the actual number of sheep in the shire was 43,010.

Scotland is not a great wheat-growing country; the summers are too wet and cold. In fact, in several of the counties not a single acre of land is given to wheat. By far the most important crop is oats, which is peculiarly well suited to our moist, cool climate. In Renfrewshire, for example, oats occupy more than seven times the area devoted to wheat. This contrasts very markedly with some of the English counties, such as Cambridgeshire, where wheat is grown over nearly twice the extent occupied by oats. A sunny, dry summer is necessary if the best conditions for wheat-growing are to be realised, and in this respect the east of England has obvious advantages over any western county. The conditions are similar in Scotland, where of all the counties Fife grows most wheat in proportion to its area. It is twice the size of Renfrew but it grows more than six times the quantity of wheat. The actual areas in Renfrewshire devoted to oats and wheat in 1910 were 10,164 acres and 1802 acres respectively. There are no other corn crops of any importance in the shire, but of other products potatoes

and turnips are the most valuable. There are 2863 acres
under potatoes and 2088 under turnips. The area given
up to hay is of course extremely large, there being 18,876
acres thus cultivated in 1910. (See diagrams, pp. 175,
176, 177.)

12. Industries and Manufactures.

Most of the industries of the west of Scotland date
their birth to some time within the last century and a
half. Their present flourishing condition is due almost
entirely to the development of the coalfields of Lanarkshire and Ayrshire. It was not until James Watt had
improved the steam engine that coal could be obtained in
large quantities. Then came the application of steam to
manufactures, each trade reacting on and stimulating the
others. The introduction of machinery in the textile
trades during the second half of the eighteenth century
was a great factor in the Industrial Revolution, and the
west of Scotland took full advantage of the inventions
of Hargreaves, Arkwright, Crampton, and Cartwright, in
England.

One of the healthiest features of the industries of
Renfrewshire, and in fact of the whole Clyde basin, is the
great variety of manufactures, and the absence of any
exclusive specialisation in any single line. We do not
find groups of towns engaged almost entirely in the cotton
trade as in Lancashire, or in the woollen trade as in West
Yorkshire, or in the iron trade as in the "Black Country"

On the Clyde

of England. This is undoubtedly a favourable state of affairs, for it is seldom that several of the great industries are notably depressed at the same time, and sudden fluctuations from excessive prosperity to the depths of adversity are not nearly so common in Renfrewshire as in certain other great manufacturing districts. This, however, was formerly not the case. More than once in the first half of the nineteenth century there was intense misery in Paisley owing to depressions in weaving at a time when little else was done in the town.

Of all the manifold branches of industry in the west of Scotland, Renfrewshire takes a pre-eminent position in two only—ship-building and thread-making—but here she is supreme. The Clyde is the greatest ship-building centre in the whole world, and Renfrewshire takes first place among the counties bordering the river. The priority of the Clyde in the last hundred years is well illustrated by the simple statement that the first passenger steamer (the *Comet*), the first turbine passenger boat (the *King Edward*), and the first turbine ocean-liner (the *Virginian*), were all built on that river. In 1910 a tonnage running into six figures was produced by four foreign countries, the United States, Germany, Holland, and France, but the single district of the Clyde, in spite of a lock-out lasting for three months, easily outdistanced all of them. We find again the same variety marking the ship-building as marked the manufactures as a whole. The river is not confined to one or two types. Every kind of craft that floats will be found on the stocks of the Clyde, from an ocean palace to a tramp, from a racing yacht to a dredger, from a battle-ship to a Port Glasgow wind-jammer.

The geographical causes of the Clyde's pre-eminence are fairly obvious. Like the ship-building centres next in importance—the Tyne and the Tees—the Clyde estuary runs into a busy coalfield, where iron and steel working and marine engineering are staple industries, and all

Dock and Ship-building, Renfrew

necessary material can be obtained without costly transport. The human element, too, has been just as important, for the river was originally quite unsuited for ship-building. The geographical factor again is well seen in one of the sub-divisions of the industry. The beautiful estuary of the Clyde holds out so many attractions to town dwellers that it made a fleet of fast pleasure steamers an imperative necessity. Therefore it is not surprising to find that Clyde-built boats, with few exceptions, monopolise all the river and cross-channel traffic of the British Isles.

In 1910 there were built on the Clyde nearly 400,000 tons of shipping, and of this gigantic total Renfrewshire claims one-half. Among the first seven firms of that year there were four Renfrewshire companies. Port Glasgow is the chief centre, and the record yearly tonnage of the river for an individual yard is consistently made by a Port Glasgow firm, Russell and Company, the total in 1910 being almost twice that of the next yard. In the building of huge men-of-war Scotts of Greenock takes a very eminent position, a single super-Dreadnought battle-ship in 1910, a *Colossus* in more than name, lifting them into seventh place among the Clyde yards.

Although Port Glasgow and Greenock dwarf the other towns, yet a respectable total is shown by Renfrew and even by Paisley. In one branch in fact Renfrew leads the world. She builds dredgers to excavate any-thing from sand to solid rock. Practically all the important harbour and dock authorities in Britain have been her customers; and hoppers and dredgers are sent out from the royal burgh to all parts of the globe. Even Gourock, not to be behind, has its speciality in the form of motor boats. Lastly, it must not be forgotten that one of the greatest events in the history of ship-building was the launching in 1812 at Port Glasgow of the famous *Comet*, the first passenger steamer that ever sailed in Britain.

In Renfrewshire, as elsewhere, spinning and weaving have been carried on for centuries, but the yarn and the

H.M.S. Colossus

(*A battleship of the super-Dreadnought class*)

cloth were for local consumption only. In the seventeenth century the textile industry, as we know it, began in Paisley. There were 66 weavers in the town at the end of that century. Cotton was not yet worked, linen and woollen goods alone being produced. In the eighteenth century the cultivation of a patch of flax and the preparation of the fibre were recognised parts of farm work. It was spun and bleached by the women, and sold to the Paisley merchants, who retailed it to the weavers of the surrounding districts. Nannie, the witch in Burns's *Tam o' Shanter*, wore a shirt of coarse linen—a "cutty sark o' Paisley harn." There was no concentration into factories. The whir of the spinning-wheel and the clack of the loom were heard in every village of the shire, far away from the stir of any town. The invention of the spinning-jenny, the mule, and the power-loom sounded the death-knell of the hand-worker, though at first the abundance of machine-spun yarn greatly stimulated the hand-weavers. Even yet, however, in a few places in the county, as at Kilbarchan, one or two hand-loom workers may still be seen.

The story of the Paisley shawl forms one of the most interesting episodes in the industrial history of Renfrewshire. The industry was born, reached a marvellous zenith of prosperity, declined, and became extinct, within the space of 80 years. Its introduction was the result of Napoleon's expedition to Egypt. Beautiful Indian shawls were sent home by the officers both of the English and the French army. The Cashmere shawls were decorated with the most amazingly elaborate patterns worked by

hand, many of them taking years of labour to produce.
The object of the Paisley workers was to produce in the
loom the effects obtained in the Indian shawl by means
of the needle. The resulting success was the reward of
patience, skill, taste, and delicacy of touch, carried to a
point that has probably never been equalled before or
since. It is not surprising to find therefore that the
Paisley weavers of this time were an altogether exceptional
class of men. They were noted for their industry, their
intellectual strength, their cultured taste, and their love
of beauty. Their characteristic independence of judgment
was fostered by the fact that they were their own masters,
and worked when and how they liked. Radicals to a
man, they took a prominent part in the struggles for
electoral freedom in the nineteenth century.

Some of the shawls were sold for £20; and in a
single year shawls to the value of a million sterling were
made in Paisley. Then the mysterious decrees of fashion
declared that the Paisley shawl should no longer be the
mode, and its doom was sealed. Queen Victoria galvanized
the dying industry into temporary life by purchasing
17 shawls and using one at the baptism of the infant
Prince of Wales, the late King Edward. But even the
will of a queen is powerless to control economic conditions
for long, and by 1880 the industry had ceased to exist.

Paisley thread is found in every corner of the globe.
An army of people in that town are employed in the
great thread mills, and the parent factories have giant
offspring in many parts of the world. The industry
was founded by Christian Shaw, daughter of the laird of

Bargarran, who when a young girl was the chief actor in
a tragedy that sent six poor creatures to the flames on
Gallow Green in 1697. This hysterical girl, plagued by
witches in her youth, grew up a shrewd and capable
business woman. Till this time yarn in this country
could not be twisted into a thread suitable for the needle,
and the thread in use was imported from Holland.

Clark's Thread Mills, Paisley

Christian Shaw, now the wife of the minister of Johnstone,
procured a hand-twisting mill from Holland, and produced
a linen thread equal or superior to the Dutch article. It
was soon in great demand among lace-makers, and
Bargarran thread became famous. Competitors in Renfrewshire were not slow to spring up, and by 1784, 120
machines were twisting thread in Paisley. Linen thread

exclusively was made at first, but the production of a fine
cotton yarn by Lancashire machinery led to the evolution
of cotton thread, which was smoother and much cheaper,
although not so strong. In the nineteenth century two
factors contributed powerfully to cause the thread industry
to remain settled in Paisley. One of these was the
abundant supply of cheap labour due to the decay of the
shawl trade and other branches of weaving. The other
was the invention of the sewing machine, which marvellously stimulated the demand for thread. It is a
coincidence worth noting that the factories of the greatest
sewing-machine makers and the greatest thread makers in
the world are situated within a few miles of each other,
on opposite banks of the Clyde.

Cotton textiles have now completely displaced linens
in the west of Scotland. The first cotton-mill in this
country was erected at Penicuik in 1778. The second
was built at Rothesay, and then Renfrewshire took a
prominent position when mills were placed at Barrhead
and Johnstone. The selection of these places seems
strange, but it must be remembered that the coalfields
were not then the all-powerful controllers of manufacturing industries that they have since become. A good
supply of water was the essential requisite. By 1787
there were 19 cotton-spinning mills in Scotland, and of
these Renfrewshire had more than her share with four.
Sixty years later there were over 50 in the county.

It has been shown above that the origin of the textile
industry in Renfrewshire was due to causes that may be
called personal or historical, that is, the human factor was

the dominant one. But the persistence of cotton manufactures in the county affords a good illustration of purely
geographical control. Consider the distribution of the
cotton industry of Britain. In Scotland cotton goods are
made chiefly in west Lanarkshire, in west Renfrewshire,
and in Ayrshire. In England, of course, the industry is
concentrated in Lancashire. The point in common to
the situation of these centres is that they are all on the
west coast. What is the cause of this peculiar distribution?
It is sometimes stated that the reason is to be sought in
the fact that the raw cotton comes from the west. But
this reason is inadequate, for the raw wool comes into
Britain from the west also, yet most of it goes to the east.
The cause is undoubtedly a climatic one. A moist
atmosphere is necessary for the manufacture of high-grade
cotton goods, otherwise the material becomes brittle and
difficult to work. Now the average rainfall of many
parts of the west coast is double that of the east coast.
Thus the west coast possesses the valuable attributes (at
least for the cotton worker) of a high rainfall and a humid
atmosphere, and therefore the industry, by a process of
the survival of the fittest, has come to be localised entirely
in the west. A climatic control working in exactly the
opposite direction is seen in the fact that very dry conditions are desirable for glue-making. Therefore a
chemical work in Greenock engaged in making glue had
to surmount this difficulty by the erection of very costly
drying and cooling apparatus. Obviously, other things
being equal, the east coast could compete successfully in
the making of this commodity.

The textiles of Renfrewshire have always comprised the more delicate fabrics, and those requiring unusual skill in making. In the linen-weaving days fine lawns and cambrics were made, so that when cotton-weaving was introduced it was natural that the special bent of the craftsmen should lead them to the making chiefly of muslins. In the manufacture of fine silk goods too, Renfrewshire has always taken a prominent position. Paisley was the first town in Scotland to engage in the making of silk. The weaving of silk gauze became a thriving industry for a time, but in recent years the silk trade has not increased. As with the other textiles the chief materials made are the lighter fabrics such as chiffons, ties, gauzes, and handkerchiefs, and in spite even of French competition the industry succeeds in holding its own.

In the last fifty years Renfrewshire has become an engineering centre of the first importance. Practically all the larger burghs are engaged in various branches of engineering ; but Paisley, Greenock, and Johnstone are the chief centres. Greenock naturally devotes itself largely to marine engineering, while Paisley and Johnstone may be said to specialise in machine tools. Every possible kind of tool is made, from the huge planing-machines for armour plates, to the half-human automatic contrivances for turning out penholders by the gross. There are machines for punching, for shearing, for rolling, for drilling, for tapping, for bending, for cutting, for flanging, for sawing—in fine for every conceivable process known to the man in the street, and many more known only to the engineer. There are foundries capable of producing the largest castings needed for marine work, sugar-machinery of all kinds is made, plant for breweries and distilleries is erected; and in Greenock there is a huge business in pumps, steam and hand steering-gears, capstans, winches, and all kinds of deck machinery and engines.

There are neither blast furnaces nor steel works in the county. The industrial centres are situated some little distance from rich coalfields, and therefore they are at a slight disadvantage in the cruder processes of metal-working, where large quantities must be turned out at a small cost. This is a geographical law of universal application. Wherever the source of power is not in the immediate neighbourhood, we find that the amount of skill and labour expended on the articles produced is large compared with the cost of the raw material, and therefore slightly increased transport rates are insignificant in comparison with the value of the finished product. Exactly the same thing holds when the manufacturing centre is some distance from its market or from a sea-port. In this country Birmingham is perhaps the best example of the second aspect of the principle, while in Renfrewshire the specialisation in costly machine tools is an excellent illustration of the first. The market of the county, however, is close at hand in the great ship-building yards that line the Clyde.

It is impossible to enumerate all the other branches of industry carried on in the county. Only a very few of the more important can be mentioned. At Renfrew are the largest water-tube boiler works in Britain; and it has now been decided that the Clyde Trust's new dry dock,

6—2

the largest in the world, shall be constructed at Renfrew, so that the rapid development of this town as an industrial centre seems assured. Sugar-refining was commenced at Greenock in 1765, and this town is still the headquarters of the trade. The industry is but a shadow of its former self, however, not owing to any lack of enterprise or skill on the part of the manufacturers, but because of the com-

Torpedo Factory, Greenock

petition of continental beet-sugar produced under an artificial, bounty-fed system. At Greenock and Port Glasgow, saw-mills, roperies, and works for sailcloth and similar things naturally bulk largely. At Paisley, starch, corn-flour, pottery, soap, and paper are made, and Barrhead contains one of the largest sanitary engineering works in the country. An industry unique in Scotland has recently been installed at Greenock by the Admiralty, namely, the manufacture of torpedoes. The fact is interesting for it furnishes another excellent example of geographical control. The town is a convenient, industrial centre, and it lies exactly opposite the mouth of Loch Long, part of which forms an ideal stretch for testing the torpedoes.

13. Mines and Minerals.

Renfrew is not a great mining county. In this respect it lags far behind its neighbours Ayr and Lanark. The reason for this has been given in the section on Geology, where it was shown that the true Coal Measures do not occur in the shire. Coal-mining is carried on, it is true, but the pits are sporadic, and the surface gear is not an integral feature of the landscape as it is in parts of Lanark. Materially the county suffers, but there are counter-balancing gains. The sheep are not transformed to a sooty hue, nor is the vegetation blighted as in the typical "Black Country" of Scotland. Such coal-mining as exists is carried on in the eastern part of the shire, and the seams worked occur in the Carboniferous Limestone series beneath the true Coal Measures. In all, less than a thousand miners are employed, a number that sinks to insignificance beside the huge total of 55,000 for Lanark.

The coal is extracted either on the "stoop and room" system or on the "long wall" system. In the first method roads are driven through the coal and connected by cross passages, leaving pillars of coal to support the roof. The roof is afterwards propped up by timber and the coal pillars removed. This method is generally employed for thick seams. For thin seams the long wall system is preferred. As the work proceeds outwards the whole of the coal is extracted, and the "face" is thus gradually pushed out, while the waste material is stacked up to support the roof. In recent years coal-cutting machinery has been largely introduced. It is used on the long wall system for thin seams, and is often made to cut through the under-clay, thus preventing any waste of coal. Some of the machines are driven by electricity, some by compressed air. In 1908 Renfrewshire produced nearly 97,000 tons of coal, which seems a fairly large figure until we find that Lanark was responsible for seventeen millions.

With regard to iron-ore, however, the county takes quite a respectable position. It produces almost as much as its larger neighbour Lanark, although both are easily surpassed by Ayr. But among them, these three adjoining shires produce far more ironstone than all the rest of Scotland put together. In the early days of the Scottish iron industry only local ores were used. A great impetus to the mining of iron ore was given at the beginning of the nineteenth century by the discovery that the miners were rejecting, under the name of "wild coal," a valuable ore known as blackband ironstone. For many years there was no need to import foreign ores, but the advantages of pure foreign haematite for steel-making, and the gradual exhaustion of the better seams of local ironstone, have caused a great change in this respect. The output of Scottish ores has fallen off rapidly. Thus in 1881 the total production of Lanarkshire and Ayrshire was nearly two and a quarter million tons, in 1890 it was nearly three-quarters of a million, while in 1908 it was con-

Quarry Workings, Giffnock

siderably less than half a million, not a fifth part of what it was 30 years ago. Most of the imported ore is brought from Spain.

The one mineral product for which Renfrew takes first place is building-stone. In spite of the small size of the county, it surpasses every other shire in Scotland in this respect, although Dumfries makes a good second; in fact, the growing popularity of the red sandstones of the latter county would seem to threaten the premier position of Renfrew. In 1908 over 150,000 tons of sandstone were produced, Dumfries and Lanark being the only other Scottish counties that ran into six figures. The famous quarries of Giffnock have built almost three-fourths of the city of Glasgow. These sandstones occur near the top of the Carboniferous Limestone series. The beds that are quarried consist of a yellowish-white, fine-grained freestone, which shows hardly any sign of stratification, and can be worked with equal freedom in almost any direction. The material is extracted mainly by open quarrying, although considerable quantities have been taken out by underground galleries, on a "stoop and room" principle, though on a huge scale. In recent years a serious problem has arisen in large towns owing to the rapid weathering of the building stones. This, curiously enough, takes place, not on the exposed parts, but on the undersides of ledges or the sheltered portions of ornamental work. It has been attributed by some to the action of acids in the air, by others to the destructive influence of certain bacteria. The cause is not clearly understood ; it is certainly not a normal type of weathering, a conclusion that is demonstrated by the position of the decayed portions. The Giffnock sandstones on the whole resist this action better than the red Triassic sandstones of Dumfries.

Some of the bands of limestone in Renfrew are of considerable economic importance, and in several places quarrying is carried on. The chief bands are the Orchard

limestone and the Arden limestone, both of which occur
above the Giffnock sandstones. The Arden limestone
reaches a thickness of 10 feet in places, and has been
largely quarried near Barrhead and Thornliebank, while
the Orchard limestone, though not so thick, is prized for
its valuable properties as a cement limestone. In 1908
nearly 20,000 tons of limestone were produced in Renfrew,
a total unequalled by any other county in the west,
although surpassed by Edinburgh and Banff.

All over the shire the volcanic rocks are used for road-
metal. On almost every hillside gashes may be seen on
the rock face, indicating where man is assisting nature in
her endless task of dragging all things to a common level.
The ideal road-metal should be hard, tough, not given to
mud-making, should offer a good footing in all weathers,
and bind well together. The igneous rocks of the county
possess all these qualities in a high degree. From the flat
lands in the east of the shire clay is obtained for brick-
making. Only three other counties in Scotland surpass
Renfrew in the production of clay and brick-earth, the
total obtained in 1908 being nearly 35,000 tons.

14. Shipping and Trade.

The geographical advantages of Renfrewshire for
shipping are obvious. The county borders the Clyde
from the point at which the river becomes navigable
down to the open firth. It is but natural therefore to
find ports of considerable importance dotted along the

river frontage. The destinies of the county have been profoundly modified by the deepening of the Clyde. Shipping almost inevitably goes to the head of navigation of a river; and when, in addition, the greatest city of the district is situated there, the tendency for shipping to be drawn into its vortex of trade is irresistible. It would be interesting, though serving no practical purpose, to specu-late on the destinies of Greenock or Port Glasgow if the Clyde had remained undredged. Certainly their shipping trade would have been much greater, though doubtless they would have lacked something of the general industrial stimulus that radiates out from every great manufacturing centre.

From Glasgow to the firth, the Clyde is largely the product of man. The conversion of a stream, in places but a few inches in depth, into a water-way for ocean-going ships is one of the romances of industrial history. In the sixteenth century an attempt was made to improve the channel at Dumbuck but was not successful. The ma-gistrates of Glasgow therefore reported in 1668 that they had had "ane meeting yeasternight with the lairds, elder and younger, of Newark, and that they had spoke with them anent the taking of ane piece of land of theirs in feu, for loadning and livering of their ships there, anchor-ing and building ane harbour there, and that the said lairds had subscryvit a contract of feu this morning; quhilk was all allowed and approvine be said magistratis and counsell."

On the ground thus purchased the magistrates laid out the town of Port Glasgow with harbours and a graving dock. Here the goods were taken from the ships and loaded on the backs of little pack-horses that brought them by badly made tracks to Glasgow. In 1755 the river was still in a state of nature, for between Glasgow and Renfrew there were twelve shoals, one of which was only 15 inches deep at low water. James Watt surveyed the river in 1769, and reported a depth of 14 inches at Hirst Ford during low water. To John Golborne of Chester is due the first marked improvement in the navigation of the river, which was dredged and also narrowed by the construction of jetties. A few years later Golborne deepened Dumbuck Ford to a depth of 7 feet, and owing to the scour of the river due to his jetty system, this depth was in 1781 found to have become 14 feet.

Act after act was carried through parliament giving new powers, and each meant a further improvement in navigation and a consequent stimulus to the commerce of Glasgow. A great advance was made by the application of steam power to dredgers, and the adoption of steam hopper barges, to which the present state of the river is largely due. A formidable obstacle was found in the Elderslie Rock, extending right across the river at a depth of 8 feet below low water. After years of labour this was removed at a total cost of about £140,000, giving now a depth at low water of 28 feet.

After its foundation Port Glasgow rapidly grew in size and importance till in 1710 it was made the chief customhouse port on the Clyde. Still the shipping grew, and in the middle of the century a graving-dock was constructed, the first ever built in Scotland. By the end of the eighteenth century, however, Glasgow was beginning to draw the over-sea traffic to itself up the deepened river, until towards the middle of the nineteenth century it almost seemed as if Port Glasgow were doomed to extinction. But the natural advantages of the town's position asserted themselves, supported by the energy and enterprise of the inhabitants, and the town, no longer bound to the chariot of Glasgow, entered on a new lease of independent life. Ship-building was vigorously started, raw material was imported, and various industries, largely with a salt-water flavour, sprang up in the town. At the present time the trade is chiefly with Canada and the West Indies, great quantities of timber being imported from the former country.

From the beginning of its history Greenock has been identified with the shipping-trade. We have already mentioned that it became a sea-port of considerable standing by the enlightened policy of its lairds, Sir John Shaw and his son of the same name. In the middle of the seventeenth century we find the inhabitants of Greenock referred to as "all seamen or fishermen trading for Ireland or the Isles in open boats." Sir John Shaw's whole life was a determined and successful struggle to make his town a sea-port of prime importance. At this time the town was fettered by its inability to engage in foreign trade. This right was the prerogative of the royal burghs, and jealously they guarded it. They kept especially a wary eye on Greenock, for they could not fail to see its natural advantages, and several times Glasgow, Dumbarton, and Renfrew tried to thrust the rising port back to the

Off Greenock: Outward Bound

obscurity from which it had sprung. But without success, for Shaw of Greenock in return for special service to the king was able to throw open the gates of foreign commerce to the town.

In 1670 we find King Charles II connected with the town, and this time in a new *rôle* even for that versatile and accomplished monarch. He became a herring-curer, doubtless with little practical knowledge of the mysteries of that art, but at any rate as a shareholder in a herring-curing company that had one of its principal stations at Greenock. Greenock was the chief centre of the herring-fishing at this time, and exported yearly thousands of barrels to the Continent. By 1728 there were 900 boats engaged in the trade.

The year 1710 was a red-letter one for the town. The harbour and quays, "most commodious, safe, and good," built by the town on the suggestion of Sir John Shaw, were finished that year amid general rejoicing, and Greenock took its place as "the chief town upon the coast, well built, consisting chiefly of one principal street, about a quarter of a mile in length." But the rising importance of Greenock roused the animosity of more powerful opponents than even its jealous neighbours, the royal burghs. The great English sea-ports, London, Liverpool, and Bristol, saw with unconcealed anger much of their valuable foreign trade being captured by this upstart Renfrewshire town. Eager to damage Greenock at any cost, they accused the merchants of being in collusion with the customs officials, and thus unfairly securing trade at their expense, and even obtained the introduction

Greenock from Whin Hill

in parliament of a bill to deprive the town of its trading privileges. Almost exactly the same charge was made against Glasgow, but it is satisfactory to know that the Scottish towns were wholly exonerated by the commission

Renfrew Ferry

that investigated the matter. The passing of the bill would have been the death-blow to the prosperity of Greenock; but luckily the strenuous representations of the Scottish members averted the danger.

During the last hundred years the most important

landmarks in the history of Greenock shipping have been
the construction of graving-docks and other harbour works.
Most of these are situated at Cartsdyke, formerly a rival,
but now swallowed up by Greenock. The first graving-
dock was built in 1786 and admitted vessels drawing
10 feet of water. The Custom House Quay, the East
India Dock, the Victoria Harbour, the Albert Harbour,
the Princes Pier, the Garvel Dock, and the great James
Watt Dock for ships of over 30 feet draught, these form
a succession of harbour works of which any town might
justly be proud.

At one time Renfrew was the principal port on the
Clyde, but its glory has departed. The causes of this
are largely geographical. Lying between Glasgow and
Greenock, it has the advantages of neither, for both
Glasgow and Greenock are at "critical points" on the
river. Greenock is at the upper end of the firth, it is
the first town to be reached on the Clyde proper, while
Glasgow has the immense advantage of being the farthest
up point on the river to which ocean-going ships can
come. This is the supreme advantage in maritime commerce; the head of ocean navigation is *the* commanding
situation—witness not only Glasgow but also Liverpool,
London, Montreal, Rangoon, and many other towns.
In addition, it is profoundly true in geography as in other
things that "Unto him that hath shall be given." The
fact that a town is large, is enough in itself to establish
a considerable ebb and flow of trade, which tends con-
tinually to direct minor streams of traffic into the main
currents.

In spite of its situation on a small and tributary river Paisley has had for some time a fair shipping trade. More than a hundred years ago the Cart was deepened and rendered navigable for boats drawing not less than five feet of water. Many difficulties have been surmounted until the town can now be reached by ships drawing 15 feet of water. It should not be forgotten that at the middle of last century the Cart was the district noted for the building of the swiftest river steamers then on the Clyde.

For many years the river steamers of the Clyde have formed a class by themselves. Dwellers on Clydeside glory in the thought that their old boats, unable to hold their own against their younger and speedier rivals, are sent to other ports, there to become the cynosure of the district, the pride of the estuary, as the last word in river craft. And, flattering though the idea is, there is considerable truth in it. There is perhaps no other place on the face of the earth where such a combination of speed, comfort, elegance, and economy in steamboat travelling can be had as on the Firth of Clyde. One can sail for a day through some of the finest scenery in Britain for ninepence! In recent years the three great railway companies of the west of Scotland have taken the lion's share of river travelling. Two of the companies have their shipping headquarters in Renfrewshire, the Caledonian at Gourock, and the Glasgow and South-Western at Greenock. From each of these ports there plies a fleet of powerful and speedy steamers, dashing back and forward between all the little pleasure towns of the firth. The first glimpse of these dainty craft, with their distinctive colour-note in hull and smoke-stack, rolls the burden of care from the shoulders of the worried business man, and brings a transient gleam of sunshine into the black lives of children pent up, save for one week in the year, in a dreary wilderness of stone and lime. Greenock, Gourock, and Craigendoran have filched from the Glasgow Broomielaw much of its ancient glory. Nowadays almost everyone prefers to journey to the beginning of the estuary by train; although quite recently a cleaner Clyde has made some prefer the slow but interesting sail from Glasgow.

15. History of the County.

The early history of Renfrewshire is largely a matter of conjecture. Two thousand years ago the district was inhabited by a tribe called the Damnonii. They are usually referred to as Celts, but we have already indicated the probability that Celtic blood may not have been nearly so prominent in Scotland as Celtic speech and culture. The county lay somewhat off the main line of the Roman invasion, so that remains of the Roman occupation are not nearly so numerous as they are for example in Lanarkshire, through the heart of which ran one of the main Roman roads. Some coins and vases, however, have been found, and there was a camp at Paisley, this town being generally identified with the Roman station of Vanduara, although Skene maintains that this was at Loudoun in the Irvine valley. The mists veiling the history of the county in Roman times are thick enough, but after the withdrawal of the legions they settle down still blacker. Almost all we know with certainty is that the Britons took possession of the district again, and it formed part of the kingdom of Strathclyde, the capital of which was Alclwyd or Dunbreatan (hill of the Britons), now known as Dumbarton.

Christianity had been introduced in the time of the Romans, but after their withdrawal the inhabitants lapsed largely into Paganism. In the sixth century, however, a number of monks from Ireland settled in the county and founded churches. Among them was the good St Mirin who laboured long and faithfully at Paisley. St Barchan and St Fillan, the patron saints of Kilbarchan and Kilallan, were the best known of the other monks. The arm of St Fillan was carried in a case at the Battle of Bannockburn, and this holy relic of the Renfrewshire monk helped to win the victory for the Scots. Another saint was given the credit of the great victory over Somerled, Lord of the Isles, that took place at Renfrew in 1164. Several years before, Somerled had devastated Glasgow, and the bishop had prayed earnestly to St Mungo to hurl divine vengeance on the spoiler's head. Somerled landed at Renfrew and marched half-way to Paisley, where he was slain and his army totally defeated.

Renfrewshire was the ancestral home of the founders of the Royal House of Stewart, whose blood runs in the veins of our present king. When David I returned to Scotland after helping Maud in the Barons' War he was accompanied by Walter FitzAlan, who was made High Steward of Scotland, and was given lands by the king, which included almost the whole of Renfrewshire. This was in accordance with the whole feudal policy of David, who had been imbued with Norman ideas and culture at the court of Henry I. Therefore we find during his

David I and Malcolm IV
(From the *Kelso Charter*)

reign an influx of Normans into Scotland, who soon settled down in permanent residence, and founded some of the most powerful families in the country. For example, Walter gave grants of land to his own friends —Eaglesham to Robert de Montgomery, ancestor of the Earls of Eglinton; part of Mearns to Herbert de

Maxwell; and Nether Pollok to John de Maxwell, all well known and honoured names in the west of Scotland.

Walter the Steward brought with him from England thirteen monks who were settled in Paisley and were given lands and houses, their possessions extending as far as Prestwick. In the twelfth century the monastery became the powerful Abbey of Paisley, which dominated the religious life of the county until the Reformation. Walter's descendants became hereditary Stewards of the kingdom, and we find their names connected henceforth with practically all the great events of Scottish history. A band of Paisley men under Alexander Stewart, great grandson of the first Walter, came on the scene at a critical time, and helped to gain the glorious victory of the Battle of Largs over King Haco of Norway. James the High Steward, son of Alexander, made an alliance with Robert the Bruce, Earl of Carrick, and staunchly supported his claims to the throne of Scotland. His name, however, stands first on the Ragman Roll, which contains the names of all those who swore fealty to Edward I after the submission of Balliol. His son Walter, the sixth Steward, was an ardent and powerful supporter of the cause of the younger Bruce, and distinguished himself by his bravery at Bannockburn. His services were rewarded by the hand of the Princess Marjory, daughter of King Robert the Bruce, and their son ascended the throne as Robert II. Walter shared with Douglas the regency of Scotland when Bruce was campaigning in Ireland, and until his death was ever to the fore against the "auld enemy" across the border.

Walter's son (the future king) was a worthy successor. In his young days he fought gallantly against the hereditary foe. In the campaign against Edward III he took a prominent part, and was one of the leaders at the disastrous battle of Halidon Hill in 1333. He was an elderly man when the death of David II without heirs gave him the crown of Scotland. Though brave and kindly, he was not energetic enough to keep the curb on the fierce and turbulent Scottish nobles. To this day the heir to the British throne claims the titles of Steward of Scotland and Baron of Renfrew.

Scotland's greatest patriot was a native of Renfrewshire. Sir William Wallace was born at Elderslie near Paisley, and the house in which he was born is still pointed out. Most of the building is of course of much later date, but part of the massive basement may possibly have been in existence when the great warrior was born. Until fifty years ago there stood near the building a mighty tree known as Wallace's oak, and tradition says that in its branches Wallace and three hundred men hid from the English. The family was an Ayrshire one, Riccarton having been founded by Richard Wallace, and thus we find that many of Wallace's adventures are associated with the Irvine valley and other parts of Ayrshire, while few are recorded as having occurred in Renfrew.

One of the descendants of Robert de Montgomery, who received from Walter the first Steward a grant of lands at Eaglesham, is famous for the part he played at Otterburn, where he captured Percy the leader of the English :

"I will not yield to a braken bush
    Nor yet will I yield to a brier,
    But I would yield to Earl Douglas,
    Or Sir Hugh the Montgomery, if he were here.

    As soon as he knew it was Montgomery
    He stuck his sword's point in the gronde;
    The Montgomery was a courteous knight
    And quickly took him by the honde."

There are variations of the ballad giving different readings, one stating that Sir Hugh was slain, another that he
was taken prisoner. It seems probable that Sir Hugh
was slain but that his father Sir John was the captor of
Hotspur. Sir John Montgomery acquired the estates
of Eglinton and Ardrossan by marriage, and thus from
him, but through a female branch, is descended the present Earl of Eglinton. The connection of the Royal
House of Britain with Renfrew has been already explained,
but the county touches the kingly line at another point,
again through the great Stewart family. The third son
of the second Walter, High Steward of Scotland, married
the daughter of Robert de Croc, Lord of Crocstoun
(Crookston) and Darnley, and founded the Stewart family
of Darnley in Renfrewshire. The descendants became
Earls of Lennox, and one of the family, Lord Darnley,
was the husband of Mary Queen of Scots, and the father
of James VI.

    After the War of Independence none of the main
movements in Scottish history took place on Renfrewshire
ground until the time of Queen Mary. The blow that
finally crushed the unfortunate queen was the Battle of

Langside, which was fought in Renfrewshire, although the ever-widening circles of Glasgow's administration have now embraced this area in the boundaries of the city. When Queen Mary escaped from Loch Leven her supporters assembled in force at Hamilton. The Regent Murray was encamped at Glasgow to prevent the passage of the Clyde, if Mary, as was expected, should move on Dumbarton. Learning that the Queen's army would attempt the passage lower down the river, the regent moved out of the town to Langside Hill in order to intercept the enemy. Here he was attacked by the Queen's forces but completely defeated them. Mary was watching the battle from a hill near Cathcart, and on seeing the flight of her army galloped off in terror, and did not draw rein till she reached Sanquhar, 60 miles away.

At the Reformation two of the protagonists were Renfrewshire men. The Earl of Glencairn struggled hard on the side of England and the reformed religion, while the Abbot of Paisley, afterwards Archbishop of St Andrews, was one of the most powerful supporters of the French party and the old doctrine. In 1561 Glencairn and a rabble of so-called Reformers committed the unpardonable crime of wrecking the beautiful Abbey of Paisley. They partly demolished the buildings, they shattered the altars, they destroyed or plundered the "mony gud jowellis and clathis of gold, silver, and silk, and mony gud bukis,...and the staitliest tabernakle that wes in al Skotland and the maist costlie."

In Covenanting times Renfrew strongly resisted

episcopacy, although the county did not take such a prominent part in the struggle as Ayr or Lanark. A detach-
ment of Highlanders was quartered in the shire to overawe
the people, which they did to perfection. "Spoil the
Philistines" was the creed of the Highlanders, and
faithfully did they translate it into works. In Cleland's
words :

> "If any dare refuse to give them,
>   They durk them, strips them, and so leaves them."

The list of those who refused to conform to the decree
of Charles included the names of all the ministers in
Renfrewshire save one. They were forced to leave
their houses when winter was at hand, and suffered many
privations, yet they kept the Covenanting spirit burning
fiercely in the county. In 1684 there took place at Paisley
a wedding of more than ordinary interest. Graham of
Claverhouse was married to Lady Jane Cochrane, and
immediately after the ceremony, he summoned his troopers
and rode out over the hills in pursuit of the Covenanters.
The desolate, wind-swept moorlands round Loch Goin
gave an asylum to many a weary hunted martyr to religion, fleeing from the sabres of the relentless dragoons
of Claverhouse. Two victims rest in the old churchyard
of Eaglesham, butchered in 1685.

In this same year occurred the unhappy rebellion of
Argyll, which came to a tragic conclusion on the banks
of Clyde. While Monmouth was attacking James II in
England, Argyll raised the standard of revolt in Scotland,
hoping that the Scottish presbyterians would support him.

Dissensions and delay led to a general break-up of his troops in Dumbartonshire. The earl crossed the Clyde and was making his way to Renfrew. He had just forded the Cart when he was attacked by two of the king's men, who concluded, in spite of his peasant's disguise, that he must be some one of position from the indifferent way in which he abandoned his horse at the river. After a struggle he was wounded and taken prisoner. In the Blythswood grounds there still exists a large block of sandstone on which the doomed earl leaned after his capture. Tradition maintains that some reddish veins in the stone were caused by the blood that flowed from his wounds.

In the seventeenth century Renfrewshire was notorious for the number of witches that seem to have chosen the place as their home. The county rose to a bad eminence about the middle of the century, when Mary Lamont, a girl not out of her teens, was accused of having sinister designs on the Kempoch Stone, and of being in league with the devil to achieve them. The stone was to be cast into the sea, because of this storms would follow, and the final aim of wrecking boats would be accomplished. The unhappy girl was condemned to death, and along with some other so-called witches of Greenock and Gourock was burned at the stake. Towards the end of the century a Mr Blackwell became minister of Paisley, and obtained great celebrity as a witch finder. He seems to have been as expert as the famous witch-smellers of the Zulus, although the heathen had the advantage of the Christians in their more merciful method of murdering the victims. Mr Blackwell either stamped out the witches or took them with him when he left the county, for, strange to say, when he departed, "the great rage of Satan in this corner of the land" gradually disappeared.

The most notorious case was that of Christian Shaw, daughter of the laird of Bargarran, and founder of the thread industry of Renfrewshire. This "poor, afflicted damsel" was but ten years of age when the effects of witchcraft became manifest. She cried out in her sleep, "and then suddenly got up and did fly over the top of a bed where she lay to the great astonishment of her parents and others in the room." Subsequently also she floated through the room and the hall, and down a long winding staircase, which naturally perplexed the onlookers still more. She was sometimes blind, sometimes deaf and dumb, "her body was often so bent and stiff that she stood like a bow on her feet and neck at once"; again, "several weeks by past she hath disgorged a considerable quantity of hair, folded-up straw, unclean hay, wild foule feathers, with divers kinds of bones of fowls and others, together with a number of coal cinders, burning hot candle grease, gravel stones, etc." Several people were accused by her of being the cause of her torments, a Commission of Enquiry was appointed to investigate the case, and no fewer than seven persons were found guilty of witchcraft. One committed suicide in prison; the other six were burned on Gallow Green, Paisley, in the year 1697. It is amazing to think that such things could happen only two hundred years ago. The hysterical girl can hardly be held responsible, but no words are strong enough to describe the conduct of men of responsible position and good education who allowed their bigotry and superstition to hurry them into acts of such cruelty.

James VI of Scotland was connected with Renfrewshire both on his father's and his mother's side. His mother, Mary Queen of Scots, was a direct descendant of Walter the first Steward through King Robert II.

Dumbarton Castle

His father Henry, Lord Darnley, was the eldest son of the Earl of Lennox, who was descended from the Stewarts of Darnley. In 1575 we find King James VI paying a special compliment to a son of the shire. One of the Crawfurds of Jordanhill had captured Dumbarton Castle and had otherwise distinguished himself in the service of the king. He was rewarded by a letter very characteristic of the monarch in whom simplicity and cunning were so curiously mingled:

> Capten Crawfurd,
>
> I have heard sic report of your guid service done to me from the beginning of the wars among my onfriends, as I shal sum day remember the same, God willing, to your greit contentment. In the mein quhyle be of guid comfort, and reserve you to that time with patience, being assured of my favour. Fareweil.
>
> Your guid friend,
James Rex.

It is not known whether Captain Crawfurd's patience was equal to the task. The gallant gentleman, however, distinguished himself later as Provost of Glasgow, and built a great part of the bridge then erected over the Kelvin at Partick. In 1800 the estate of the Crawfurds came into the hands of the Smiths of Jordanhill, several members of which family have shed distinction on their native county.

At the rebellions of the Fifteen and the Forty-five, Renfrew like all the other south-western shires was a staunch supporter of the House of Hanover. The people had suffered too much from the Stewarts to look on their return with pleasure. In his retreat from England the Young Pretender marched up Annandale and down the Clyde to Glasgow. The Provost of Glasgow maintained that his only recruit from that town was "ane drunken shoemaker." He fared little better in Renfrewshire, and Paisley and Renfrew were among the towns on which fines were imposed.

During the nineteenth century Renfrewshire was prominent in the agitation for social reform. At the time of the Chartist movement and the Reform Bills, riots broke out in various parts of the shire, particularly in Paisley and Greenock. In those days the terms "weaver" and "radical" were synonymous. At the same time as they wove new and striking textures on their looms, the warp and woof of their minds seemed irresistibly to become fashioned on a bolder and more daring pattern than that of other workers. But it must not be forgotten that these men were intellectually of a high order. They could not but see that in intelligence, in industry, in skill of hand and eye, in cultivated taste, in love of literature and science, they were far superior to many possessing social rights denied to them. Apart from slight disturbances due to politics or to industrial depression, the history of the county since the '45 has been one of uninterrupted peace and progress. Its famous events have been discoveries in industries, its revolutions have been those of manufacturing methods, and the improvement in social conditions and customs has been no less marked.

16. Antiquities.

The earliest men in Britain were unacquainted with the use of metals. Their weapons and tools were of stone roughly shaped and chipped. These weapons of Palaeolithic type do not occur in Scotland; but stone weapons and tools, of the Neolithic type, finely chipped or polished, have been discovered in many parts of the country. They consist of celts or axes, arrow-heads, spear-heads, flail-stones, knives or scrapers, slick stones for softening hides, and other implements. Many of the so-called Druidical monuments were probably erected by Neolithic man. The race was widely distributed, stone structures of a similar kind having been found all over Europe, in Africa, in Asia, and in America. The cromlechs (or table-stones) and menhirs (or standing stones) of Scotland probably belong to this period. The Neolithic race had long skulls, good features, dark complexions, and were about 5 feet 4 inches in average height. There are one or two standing stones in Renfrewshire but no good examples of cromlechs or of stone circles. South of Kilbarchan there is a curious stone known locally as the Clochodrich or Stone of the Druid. Its significance is unknown although its name would suggest that it was associated with Pagan rites, or possibly, as Chalmers suggests, it was used by the natives of Strathclyde as a base for signal fires in times of battle. Near Langside a good specimen of "cup and ring" marking was discovered. The purpose of these curious carvings is not known, but they are found all over Scotland, often in the most inaccessible spots.

As man progressed in civilisation the art of metal working was discovered, and the earliest metal implements are made of bronze because of its easy working. These at first imitated the stone tools, so that we find celts and other implements fashioned like the stone ones but of bronze. With the introduction of bronze another race is associated in Britain. These immigrants were much taller and stronger than the long-headed Neolithic race, they had fair complexions, and their skulls were round in shape. To the Bronze Age and perhaps earlier belong the "pile dwellings" found in different parts of the country and called crannogs in Scotland and Ireland. Many of the British crannogs, however, are of much later date than the Bronze Age, some perhaps being as recent as the ninth and tenth centuries of our era. Much controversy has raged round the crannog discovered some years ago at Dumbuck on the north of the Clyde, but in 1901 a crannog was found in Renfrewshire which fortunately has given rise to no scientific bickering. This pile-dwelling occurs at Langbank, and at high water the existing platform is covered to a depth of two or three feet. From the crannog was obtained a highly ornamented bone comb which throws perhaps a little light on the age of the structure, for a very similar comb was found in East Lothian associated with late Roman relics. In addition to several other articles, there were found a small bronze fibula or buckle, and drawings made on shale. A few years ago (in 1907) an interesting and important discovery was made near Cathcart. This was a burial place belonging to the Bronze Age, in which no fewer than eight interments had been made. The bodies had first been cremated.

Few authentic traces of Roman occupation are to be found in Renfrew. Dunlop in his history describes a large Roman camp at Paisley. The street known as

Causewayside in Paisley is said to take its name from a Roman road on that site, and Paisley itself is generally identified with the Roman station of Vanduara. In Cathcart parish a Roman vase was found which is now in the Glasgow Hunterian Museum.

Near Barochan on a fine situation on the summit of a little hill stands the most interesting memorial stone in Renfrewshire. This is the Barochan Cross, beautifully proportioned and finely carved. Semple tells us that it is "a Danish stone, being full of wreathed work, such as lions and other wild beasts; but no letters are legible." History is silent regarding the stone, but local tradition ascribes it to a defeat sustained by the Danes in the neighbourhood. From the pattern of the carving it has been considered to date from the tenth century. Many stone coffins containing human bones have been discovered in the locality. Another so-called Danish stone may be seen near the Paisley Waterworks. In all probability it is the broken remnant of a religious cross such as may well have been set up on the road leading to such a popular shrine as that of St Mirin. Semple tells us that it was in his time (the latter half of the eighteenth century) that the cross piece on the top was broken off.

The modern parish church of Renfrew is believed to stand on the site of the original church granted by David I to Glasgow. The church contains an interesting monument dating back to the seventeenth century. It is a tomb erected to the memory of Sir John Ross and his wife, Dame Marjory Mure. This Sir John is said to have been created by James IV the first Lord Ross of

Effigy of Sir John Ross of Hawkhead in
Renfrew Parish Church

8—2

Hawkhead. Effigies of the knight and his lady lie on a sculptured tomb enclosed within an arched canopy, the front of which is richly ornamented with vigorous carving. Sir John is known traditionally as Palm-my-Arm from his wrestling encounter with a famous English champion. The Scot, outmatched in size and strength, acted at first on the defensive, and the Englishman was unable to seize him, particularly as his body was well oiled. The English champion then tempted Ross by stretching out his arms with the invitation "Palm my arm." The Scot seized his wrists, jerked his shoulders from their sockets, and dispatched him.

17. Architecture—(*a*) Ecclesiastical.

The earliest Celtic examples of ecclesiastical architecture were dry-built stone cells with a roof closed with overlapping stones and flag-stones. These were followed by the Columban Scottish churches, consisting of one small oblong chamber with one door and one window. No ornamentation was used until the Romanesque influence made itself felt, introduced by the Normans. The type was elaborated later by the addition of a chancel.

The Celtic structures were superseded by churches of Norman style, introduced in the twelfth century. This style is characterised chiefly by simple massive forms and semicircular arches. As a rule there is little ornament except in the doorways, the arches of which are

moulded, and into which zigzag or bird's-head ornamentation is introduced. Very few good examples of
this style exist in Scotland, but parts of the cathedrals
of Dunblane and Kirkwall, and the abbey of Dunfermline
exhibit it very well.

The round Norman arch was replaced by the pointed
arch, giving the First Pointed Style, which reached
Scotland in the thirteenth century. Fresh ornamentation was introduced, showing itself in mouldings and in
vigorous foliage. The windows were always pointed,
narrow and lofty, and an effect of greater spaciousness
combined with lightness was aimed at.   In Scotland
the style was not so pure as in England or France, as
round Norman forms lingered on, especially in doorways,
although the general style was altered. This period is
well exemplified in parts of Paisley Abbey.

From the middle of the fourteenth to the middle of
the fifteenth century, the Middle Pointed or Decorated
Style prevailed in Scotland. The details aimed at a still
lighter effect. The windows were enlarged, the tracery
became more ornate, and the vaulting and buttresses were
made lighter. Perhaps the finest example of the style
in Scotland is Melrose Abbey. The nave of Glasgow
Cathedral is also a good fourteenth century example of
the Decorated Style.

The transition to the Third Pointed Style was gradual.
In England the tracery became more rigid, and the
windows were carried up in straight lines so that the
style was called Perpendicular. In Scotland the exterior
is generally marked by rather heavy buttresses, terminating

in small pinnacles.   The semicircular arch is often used
and there is a revival of early ornamentation.   Most of
the examples are not cathedrals but collegiate churches.

There is very little undoubted Norman work in
Renfrewshire.   Part of the masonry of the old church
of St Fillan's may date back as far as Norman times,

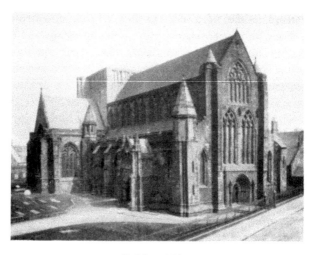

Paisley Abbey

although most of the church was certainly built at a
much later period.   Architecturally the reputation of
the county rests mainly on the beautiful Abbey of Paisley.
It was founded in the twelfth century by Walter the first
Steward, who brought monks from Shropshire, his native
county, and settled them first in Renfrew and afterwards
in Paisley.   The monastery, created an abbey in the

thirteenth century, was so enriched by the generosity of the Stewarts that it became one of the wealthiest and most powerful houses in Scotland. The first buildings were destroyed by the English seven years before the Battle of Bannockburn. Most of the existing structure was built in the fifteenth century by Abbot Thomas Tervas and Abbot George Shaw. The former "wes ane richt gud man. The body of the kirk fra the bucht stair up he biggit, and put on the ruf and theekit it with sclats, and riggit it with stane." Abbot George Shaw added to the buildings, and in addition surrounded the Abbey and grounds with a fine stone wall, which remained almost entire, till in 1781 the Earl of Abercorn sold the stones for house building. A stone still exists on which may be read the inscription :

> Thei callit ye Abbot Georg of Schawe,
> About yis Abbay gart mak yis waw;
> A thousand four hundreth 3heyr
> Auchty and fyve the date hit ueir.
>> Pray for his salvatioun
>> That made this noble fundacioun.

The attempts to erect a tower on the Abbey were unfortunate. The first tower fell, probably because of its insecure foundations. Abbot Hamilton rebuilt it at great expense, but this tower too came crashing down, destroying in its fall the choir of the Abbey. According to one account it was struck by lightning, but Hamilton of Wishaw states that it fell by its own weight. We have already seen that the Reformation brought disaster to the old Abbey. In 1557 a body of Reformers drove out the monks, sacked the building, and "burnt all the ymages and ydols and popish stuff in the same."

Most of the Abbey belongs to the Late Pointed Period. The east part of the south side of the nave, however, is undoubtedly older. The doorway here is in the Transition style, and exhibits the persistence of Norman influence. The arch head is not pointed but rounded, and has numerous bold mouldings. Above this door are three simple pointed windows which seem to be Early First Pointed work. This part of the building probably dates from the first half of the thirteenth century. To this century also must be ascribed the western doorway, a single pointed and deeply recessed opening. The upper portion of this part of the building, however, is certainly of later date, and was probably added when the abbey was restored in the fifteenth century. The north and the south transepts are in ruins, but the north wall exists with a fine traceried window which has recently been restored. The design of the triforium is very remarkable. It consists of large segmental arches which spring from clusters of piers introduced between them. To some extent it resembles the triforium of the nave of Dunkeld Cathedral, but the Paisley work is superior in style and seems of earlier date.

The Chapel of St Mirin is on the south side and is known also as the Sounding Aisle. It gets this name from the wonderful echo, which is described in great detail by Pennant in his *Tour*. "The echo is the finest in the world. When the end door is gently shut the noise is equal to a loud peal of thunder. If a good voice sings, or a musical instrument is well played on, the effect is inexpressibly fascinating, and almost of a celestial character. But the effect of a variety of instruments playing in concert is transcendingly enchanting, and excites such emotions in the soul as to baffle the most vivid description." It is hardly necessary to state that Pennant's command of adjectives is greater than his accuracy, or else the echo has weakened much in its old age. The chief object of interest in the chapel is "Queen Bleary's" tomb. It was found lying in fragments near the Abbey and was put together in 1817. The recumbent figure on the tomb is believed to represent Marjory, the only daughter of Robert the Bruce, and the mother of Robert II, who was killed by a fall from her horse between Paisley and Renfrew. Another person prominent in history, to whom a memorial is erected in the Abbey, is Sir Alan Cathcart, a faithful companion of the Bruce, and one of the knights who set sail for the Holy Land with the heart of the king. The Cathcart pillar, one of the south piers of the nave, bears the arms of the house of Cathcart in memory probably of Sir Alan.

The Abbey remained in a wretched state of disrepair until the second half of the nineteenth century. " In 1859 a more dreary place of worship it was impossible to conceive. It was like a charnel house. The burial ground outside reached above the sill of the windows. The floor was earthen and you were afraid if you stirred your foot you would rake up some old bones that lay uncomfortably near the surface." Since that time the

church has been greatly restored, and at the present time
it is one of the finest specimens of ecclesiastical architecture in the west of Scotland. The part of the Abbey
now in use contains a number of fine stained glass
windows.

The Abbey of Paisley dwarfs in comparison the
importance of all the other ecclesiastical buildings of
Renfrewshire, but some of the others present features
of interest. The Castle Semple Collegiate Church is
remarkable in some respects. The building stands on
gently sloping ground overlooking the calm waters of
Castle Semple Loch. The first Lord Sempill founded
a collegiate church on this spot in 1504. The style
of the east end of the church is very unusual. It contains double windows, the forms of which indicate that
they are late survivals of spurious Gothic work. The
style of the east end seems to stamp it as a later sixteenth
century addition. The square tower at the west end is
extremely simple and has no analogy with the work at
the other end of the church. The church contains a
large monument to Lord Sempill, who fell at the Battle
of Flodden. The style of the tomb shows the influence
of Renaissance forms on the earlier Gothic. This is
well exhibited in the foliage of the upper part which
is luxuriant to the point of exaggeration.

Between Houston and Kilmacolm in the rich valley
of Strathgryfe stand the ruins of the old church of St
Fillan's. The walls are fairly well preserved but the
gables are in ruins. Part of the walls may date back
as far as Norman times, but the openings show work

Auld Kirk, Lochwinnoch

of a much later period, a doorway in the south wall being dated 1635. At Kilmacolm there is part of the wall of an old church which may possibly belong to the thirteenth century. There are three plain lancet windows the work on which is certainly of an early date.

The Reformation put an end to medieval ecclesiastical architecture in Scotland. A few churches were certainly erected under the influence of the Episcopalians, but the Presbyterians attempted to eliminate everything that savoured of the old forms, and to this end were content to erect buildings that had absolutely no claim to respect so far as their architecture was concerned. In the eighteenth century, however, there arose in England a distinct revival of the interest in architecture, and particularly in classical styles. This feeling hardly stirred in Scotland till the nineteenth century. We are told that in the eighteenth century the Scottish churches "were disgraces to art and scandals to religion. They were mean, incommodious and comfortless; the earth of the graveyard often rose high above the floor of the church, so that the people required to descend several steps as to a cellar, before they got entrance by stooping into the dark, dismal, damp and hideous sanctuaries." At the beginning of the nineteenth century, however, a great change for the better began to take place. Architects made a special study of old buildings and old styles, and this combined with the rapidly increasing wealth of the country was soon reflected in many noble ecclesiastical buildings. The great and wealthy industrial communities

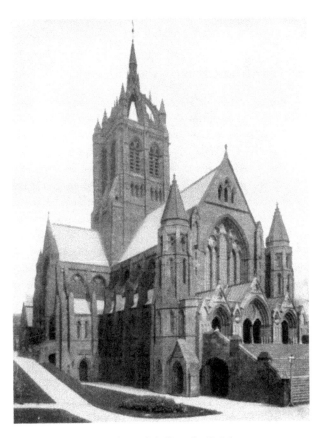

Coats Memorial Church, Paisley

of Renfrewshire can now without exception boast of
modern churches that will bear comparison with those
of medieval times.

18.   Architecture—(*b*) Military.

The finest of the old castles of Scotland were erected
in the thirteenth century. The nobles were rich, labour
was cheap, and the feud with England had not yet become
chronic. It is to this period that the magnificent pile of
Bothwell Castle belongs, with walls 60 feet in height and
over 15 feet thick, with towers, a chapel, a great hall,
and dominating everything, the grand donjon. The end
of the thirteenth century, however, marks a great change
in the style of the castles of Scotland. The War of
Independence against England completely exhausted the
resources of the country, and consequently we find that
large and massive buildings such as Bothwell Castle were
no longer erected. Their place was taken by strong,
square towers, simply fashioned after the model of the
Norman keeps. These are the castles of the second
period. They are specially characteristic of the fourteenth
century, but continued to be built at much later dates,
and from the simplicity of the design it is often difficult
to determine the exact age.

In the fifteenth century the plan was slightly elaborated. The simple, square, keep-like style was retained,
but the castle was built round a central quadrangle or
courtyard. In addition a separate tower or keep is often

found, capable of being defended although the rest of the
castle should be captured. These are the third period
castles, and most of the Renfrewshire structures are
examples of this class. All the castles built in the reign
of James I and until that of James V are of the modified
keep style. All this time, however, the defensive features
were becoming less in evidence while domestic require-
ments were demanding more consideration. Thus we
find that what was originally a necessary feature of
successful defence became later merely ornamental, and
thus also while the thick walls were retained they were
honeycombed with chambers. There are several good
examples in the county of the castles of the third period.

Life in these castles was simple to the point of rudeness.
The floors were strewn with rushes, and even in the bed-
room of the king, grass was used in place of our carpets.
The table and the side-board were the chief articles of
furniture. Forks were unknown; their introduction in
the seventeenth century was described with indignation
as "an insult to Providence, who had given us fingers."
Glass was costly and was little used. Even King Henry III
had but one glass cup which had been given him as a
present. When a noble moved from one castle to another
he carried his valuable windows with him, and set them up
in his next residence. The food if plentiful was coarse;
the manners not less so. In early times a guest who sat
in a place more honourable than his due was liable to be
pelted by the company with bones, as a hint to seek a
position more suited to his rank.

There are no ruins of castles of the first period in

Renfrew. None of the existing structures can be compared in size and magnificence with such a fine example of the first period type as Bothwell Castle. Of the castles of the second period, the simple Norman keep, Duchal Castle is the only well-known example. It was built on a rocky knoll with precipitous sides. A strong wall enclosed part of the knoll, and from one corner there rises suddenly for 20 feet a rock on which the castle was built. The structure is now a shapeless ruin, too dilapidated for satisfactory results to be obtained from its examination.

From the point of view of the antiquary or the architect, Mearns Castle is one of the most important in Scotland, for its date of erection is known exactly, and therefore it forms a standard by the help of which the age of other castles of the same kind may be approximately fixed. In 1449 James II granted a license to Herbert, Lord Maxwell, "to build a castle or fortalice on the Barony of Mearns in Renfrewshire, to surround and fortify it with walls and ditches, to strengthen it by iron gates, and to erect on the top of it all warlike apparatus necessary for its defence." Most of the castles of this style and period in Scotland are quite undated, so that authentic information of this kind is of the highest value. About the middle of the seventeenth century the castle and lands were sold by the Earl of Nithsdale (Lord Maxwell) to Sir George Maxwell of Nether Pollok, and shortly afterwards they passed into the hands of the ancestors of the present owner, Sir Hugh Shaw Stewart.

Leven Castle stands near the shore a mile or two south

of Gourock, and commands a magnificent view across the
Firth of Clyde.   It is built on the plan of a double tower,
a very unusual design for the period to which it belongs.
It seems probable that a keep of the usual fourteenth or
fifteenth century type was first erected, and later, about

Crookston Castle

the beginning of the sixteenth century, another wing was
added.   Until the middle of the sixteenth century the
lands belonged to the Mortons.   Then they passed into
the hands of the Sempills, and now they are in the
possession of the Shaw Stewarts.   A mile farther down
the firth is Inverkip Castle on the site of an old stronghold

that stood on this spot in the days of Bruce and which is mentioned by Barbour. The ruins of a square tower are all that is left of the subsequent building which was erected probably towards the end of the fifteenth century.

Crookston Castle is in many respects one of the most interesting ruins in the county. It is situated on the summit of a slope some three miles east of Paisley, and forms a conspicuous feature of the landscape. In its best days the castle must have been an imposing structure, with two lofty towers, massive walls, and battlemented wings. It is surrounded by a mound and a great ditch in a wonderfully fine state of preservation. In the twelfth century the estate belonged to Robert de Croc, one of the companions of Walter the first Steward. Later it passed into the hands of the Stewarts of Darnley, and was held by Lord Darnley, the husband of Queen Mary. Here tradition says their betrothal took place, and here they spent the days succeeding their marriage. From this spot also Queen Mary is said to have viewed the Battle of Langside :

> " And one bright female form with sword and crown,
>   Still grieves to view her banners beaten down."

Both Wilson and Scott go astray on this point, for rising ground intervenes between Crookston and Langside. In reality Mary watched the final ruin of her hopes from the summit of a knoll near Cathcart Castle. Crookston is celebrated in the verses of Burns, Tannahill, Wilson, and many other poets of lesser note.

Barr Castle south of Lochwinnoch is another of the

third period structures, and is in a fine state of preservation. In plan it is a simple parallelogram enclosing a central quadrangle. Entrance to the courtyard is obtained by a round arched doorway defended with shot-holes. For purposes offensive and defensive there are slits for arrows

Cathcart Castle

and ports for guns. Battlements extend all round the castle, and the corner turrets were originally roofed. Its date of erection was probably the beginning of the sixteenth century. Cathcart Castle was a stronghold in the days of Wallace and Bruce, but the present building is of fifteenth century style. It stands on the steep bank of the White

Cart, which defends it on two sides. It was inhabited till
about 1740 when it was partly demolished for building
material so that only one ruined ivy-mantled tower remains.
At the foot of the Gleniffer Braes there is a little lake
forming a reservoir for Paisley, and mirrored in its bright
waters is the hoary, corbelled tower of Stanely, sung by
Tannahill :

> " Keen blaws the wind o'er the braes o' Gleniffer,
>   The auld castle turrets are covered wi' snaw,
>   How changed frae the time when I met wi' my lover
>   Amang the broom bushes on Stanley green shaw."

The Danzielstons are said to have built a keep here in
the fourteenth century, but the style of the castle indicates
a later date. It was perhaps built by the Maxwells in the
fifteenth century.

With the fourth period there is a marked break in the
continuity of the style of architecture. The introduction
of Renaissance forms is obvious in the frequent use of
florid ornament in place of the Gothic severity. The
change was facilitated by the union of England and
Scotland in 1603. The use of artillery rendered an
impregnable stronghold impossible. Thus we find that
while the plan of the building remained the same, the
external appearance and the details were altered. The
transition from military to domestic needs is shown in
such a change as the evolution of angle turrets into bow
windows, a change which is typical of the whole process.
Newark Castle, Port Glasgow, belongs to the fourth period,
and is one of the finest specimens of seventeenth century
architecture in Scotland. It is built round a courtyard,

forming three sides of a quadrangle open to the south. The hall is a splendid apartment nearly 40 feet long, and is lighted by windows on both sides. The keep dates from the end of the fifteenth century, but most of the castle is more than a century younger. The doorway

Stanely Castle

shows clearly the encroaching of Renaissance details on the old Scottish design. Above it is the inscription "The blessingis of God be heirin." Haggs Castle is another fine example of the same period. It was built about the same time as Newark to which it shows many points of similarity, particularly the general richness of effect. Sir John Maxwell of Pollok was its builder, and its erection seems to have overtaxed his resources, for he wrote in 1587 to his father-in-law that his house was nearly finished but wanted furnishing "whilk is na lytell mater." He is ashamed that his home "sowld stand lyik ane twme kirne."

19. Architecture — (c) Municipal and Domestic.

Paisley can boast of finer public buildings than any other town in Renfrewshire. The Municipal Buildings are nearly a hundred years old, and form an interesting, castellated pile with projecting hexagonal turrets. The County Buildings and the Sheriff Court House are two very fine buildings in the Italian style. The most conspicuous structure in Paisley apart from the Abbey is undoubtedly the George A. Clark Town Hall, one of the most magnificent public halls in Scotland. It cost £100,000 to build, the money being given by members of the Clark family, the famous manufacturers of thread. The architecture is Greek, and from all quarters the building has a most imposing appearance. Nor is the other great thread family of Paisley unrepresented. The fine public library and museum was presented by the late Sir Peter Coats, while his brother continued the good work by erecting the well-known observatory of Paisley. Probably no other town in Britain has received such handsome benefactions from but two families. For a single edifice the Municipal Buildings of Greenock probably hold the palm for the county. They form a stately pile of Renaissance architecture, surmounted by a fine dome-capped tower. It is interesting to remember that the former town buildings on the same site were

Dunn Square and Clark Halls, Paisley

designed by the father of James Watt. The town hall of Renfrew is a pretentious building, impure Gothic in style. Although by no means unimposing, the building is not admired by architects (see p. 148).

Although many of the mansion houses of Renfrewshire are fine specimens of architecture, it may be said with

Municipal Buildings, Greenock

truth that the outstanding feature of the great estates is
the beauty of the grounds. Pollok House, the seat of
the Maxwells, for example, though internally handsome
is externally severe, not to say ugly, but the surpassing
beauty of the surroundings would redeem a much plainer
structure. Similarly Blythswood House, while handsome

Blythswood House, Renfrew

and commodious, draws its chief charm from its magni-
ficently wooded park. Hawkhead House near Paisley is
one of the most interesting mansions in the county.
Originally a tower it was greatly enlarged in the time
of Charles I, and received a visit from James VII when
Duke of York. It was repaired again in the eighteenth
century, and is now a picturesquely irregular old pile with

beautiful gardens and a fine park. Ardgowan House, the
seat of the Shaw Stewarts at Inverkip, is not only a fine
mansion, but has a magnificent position, sheltered behind
by splendid trees, in front overlooking the glorious panorama of the Firth of Clyde. Erskine House is an imposing
specimen of Tudor architecture, but it is difficult to know
whether to admire most the splendid mansion, or the finely

Old Houses at Renfrew Cross

wooded grounds surrounding it, or the rich and varied
prospect that it overlooks. One of the most picturesque
of all the mansions of Renfrewshire is Pollok Castle.
It crowns an eminence near the Balgray Reservoir, and
commands a magnificent view up and down the Clyde
valley, with Ben Lomond and other Highland giants
looming blue in the distance.

The richness of Renfrew in even-grained sandstone or freestone has led to most of the buildings in the county being constructed of that stone.   Good Giffnock stone is probably more durable than the red sandstones of south Scotland, but in recent years the imperious demands of fashion have brought about the surprising result that stone is brought all the way from Dumfries to within a mile of quarries unsurpassed in all Scotland.   There is abundant brick-clay in the low parts of the shire bordering the Clyde, but it has been little used for domestic architecture until recently.   Within the last few years, however, large numbers of cottages and small villas have been built of brick coated with rough-cast.   The volcanic rocks of the county are sometimes utilised for houses.   The result is certainly durable enough, but the stone is too hard and difficult to work ever to come into common use.

20. Communications—Past and Present.

The routes from one part of a country to another are determined by two conditions. First, if there is a demand for communication between two places, it is certain that some connecting route will be found. Secondly, the nature and details of this route will be determined by the physical features of the district.   There is no county in Scotland that illustrates better than Renfrew how the directions of routes have been controlled by the relief of the land.   In Renfrewshire the first condition for the establishment

of routes of traffic is realised by the demand for communication between Glasgow and its densely populated
neighbourhood on the one hand, and on the other the
fertile Ayrshire plain and the south of Scotland generally.
Renfrewshire lies between the Clyde and the south-western
plain of Scotland, and therefore the routes must pass

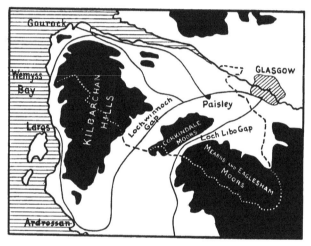

Sketch-map illustrating the control of routes by
physical features

(*Land over* 500 *feet above sea-level is shown black*)

through the shire. The actual directions taken are not
arbitrary but are rigidly controlled by natural conditions.
In order to master nature man must obey her, a paradox
that is well illustrated by the choice of routes across the
county.

In a former chapter it was shown that the high ground of Renfrewshire is divided into three great blocks with intervening valleys. This is clearly brought out by the sketch-map on p. 140 where high ground is shown in solid black. It is plain that there are two gaps in the hills, that seem created to serve as channels of communication from north to south; and through these natural doorways, the broad Lochwinnoch Gap and the narrow Loch Libo Gap, there pours a ceaseless stream of traffic by road and rail. It is plain too that along the northern and the western borders of the county there will be an easy natural route following the river and the firth. Luckily, too, for human constructions, the last geological movement in this county laid dry an old sea-beach, thus forming a narrow, flat platform that fringes the county, and that forms an ideal natural route on which to build roads and railways. There are then three great natural routes in Renfrewshire, two cutting through the heart of the hills, and one skirting them round the coast.

The Loch Libo Gap is one of the most important channels of communication in Scotland, for through it runs the main Glasgow and South-Western Railway line that connects Glasgow with Kilmarnock, Dumfries, Carlisle, and London. The railway enters the valley at Barrhead and leaves it at Caldwell. Although on both sides the hills rise nearly to 1000 feet above the sea, the summit-level of the railway is only about 400 feet above the sea. Throughout its whole course through the valley the railway runs side by side with the main road to Irvine and Kilmarnock. A Caledonian line (the

Lanarkshire and Ayrshire) runs through the valley in a parallel direction but rather higher up the slope. This line terminates at Ardrossan. There is a slightly more direct route to Kilmarnock by a gap farther east, through which runs the high road by Newton Mearns and Fenwick. The reason why this route was not adopted by the railway engineers is obvious when it is seen that any line following this direction must rise more than 700 feet above sea-level.

The easiest route across the natural barrier formed by the Renfrewshire hills is the broad, flat valley that runs from Johnstone to Dalry. We have seen that formerly this valley was occupied by a long lake or a chain of lakes, the diminished representatives of which are Castle Semple Loch and Kilbirnie Loch. Nature has formed this route so well that at the watershed the valley floor is little more than 100 feet above sea-level. Down it runs the main Glasgow and South-Western line to Ayr and Stranraer, while at Kilwinning a branch passes west to Ardrossan and then north to Largs. The railway line to Largs is perhaps the clearest example in all Scotland of the control of a route by the physical features of the district. From Paisley to Largs the distance as the crow flies is 17 miles. The distance by rail is more than twice that figure, namely 36 miles. The Kilbarchan Hills interpose a natural barrier which it is much easier to circumvent than to surmount. The line therefore goes round the southern extremity of the hills, and then runs due north to Largs. For road traffic also this route is exceedingly important.

In exactly the same way the river and coast route is used both by road and rail. The Caledonian line turns north-west at Paisley and runs by Houston and Bishopton to the river at Langbank. Thence it goes along the old raised beach to Port Glasgow, Greenock, and Gourock. Before reaching Greenock the Wemyss Bay line leaves the shore, for between Greenock and Inverkip the hills are trenched through by a deep valley that forms a con-venient short cut, and is therefore utilised by the railway. (See Fig. on p. 140.) The line ends at Wemyss Bay pier just on the county boundary.

The volcanic hills are deeply etched by the broad valley of the Gryfe, and this route is consequently used by the Glasgow and South-Western line to Greenock. The railway turns to the north-west at Elderslie, passes through Bridge of Weir, and goes up Strathgryfe to Kilmacolm. A break in the hills enables it to reach the river side above Port Glasgow, whence it runs to Princes Pier in Greenock.

In the lower parts of the county high roads cross and recross each other in every direction. Naturally hill roads are much less frequent, and the modern tendency is for the traffic to keep lower and lower down. Several of the old high roads of Renfrewshire rise over 700 feet above sea-level, but as traffic by road and rail becomes faster and faster it is found that the lowest road is ever the shortest. Although communication in Renfrewshire has been kept up for many centuries along the routes indicated, yet proper roads are of comparatively recent origin. In former times wheeled traffic was hardly possible, and most of the trade was done by pack-horses. Before the Clyde was deepened so that boats could come to Glasgow, the goods were carried from Dumbarton or Port Glasgow by pack-horses. It was not until the middle of the eighteenth century that a stage coach ran between Edinburgh and Glasgow, and the condition of the roads may be judged from the fact that its average rate was less than four miles an hour. The passing of the Turnpike Roads Act in 1751 marked the beginning of a new era, and good roads gradually replaced the old horse tracks. A further improvement was heralded by the establishment of County Councils in 1889, and the transference to them of the care of the roads. In recent years the development of fast motor traffic has presented a new problem to road authorities; and it may be that we are but at the beginning of an altogether new phase of road construction rendered necessary by the modern craze for speed.

At the present time there is no canal in use in the county, but one can still find stretches of the old Glasgow and Paisley canal. A hundred years ago it was a busy waterway carrying often a thousand passengers a day. In 1805 an act was passed for constructing a canal between Glasgow and Ardrossan, but owing to financial difficulties the canal got no farther than Johnstone. Thirty years ago the Glasgow and South-Western Railway Company obtained powers to enable them to fill up the canal and construct a railway in its place. The "Canal" station at Paisley serves to remind the passengers who rush between Paisley and Glasgow in ten minutes, of the more leisurely methods of transit of our predecessors.

The palmy days of canal traffic both for passengers and goods have passed away. As railways were extended the importance of canals declined. The complete explanation of this is by no means easy. It has been attributed to their passing into the control of railway companies, but this explanation is not satisfactory. The smallness of the vessels in use and the consequent additional handling of goods undoubtedly militate against the greater use of canals in these days, when the whole tendency is to handle and carry goods in as large amounts as possible. With the adoption of improved methods of traction or propulsion, there seems no good reason why the importance of canal traffic should not to some extent be restored.

21. Administration and Divisions.

Renfrew was originally included with Lanark as an administrative unit, the separation having been made by King Robert III at the beginning of the fifteenth century. At first the position of sheriff was a hereditary one, and was held by one of the powerful families of the county. The first sheriff that we know of was John Semple of Eliotstoun, who held office in 1426 soon after Renfrew and Lanark were separated. The office remained in the Semple family till it was transferred to the Earl of Eglinton in 1648. Until the Reformation the lands belonging to the Abbey of Paisley were not under the jurisdiction of the sheriff. The abbot was supreme, and had his gallows for hanging men, and his pit for drowning women

malefactors.   The sheriffdom was held by the Eglinton
family for a hundred years, since when appointments to the
office have been made by the crown.

Court House, Greenock

The county is divided into two wards, an upper and
a lower, the chief towns of which are Paisley and Greenock
respectively.   The upper ward contains the parishes of

Abbey Paisley, Cathcart, Eaglesham, Eastwood, Erskine, Houston, Inchinnan, Kilbarchan, Lochwinnoch, Mearns, Neilston, and Renfrew. The lower ward contains the parishes of Greenock, Inverkip, Kilmacolm, and Port Glasgow. The county possesses a lord-lieutenant, a vice-lieutenant, and a large number of deputy lieutenants and justices of the peace. The administration of the law is carried out chiefly by sheriff-courts held at Paisley and Greenock. The police force is a county constabulary, except in the large burghs, which have their own separate forces. For parliamentary purposes the county is separated into two divisions, East Renfrew and West Renfrew, each of which returns one member. In addition Paisley returns a member, Greenock another, and Port Glasgow and Renfrew have a share in a third.

County Councils were established in 1889, and look after the finances, roads and bridges, water supply, public health, and the general administration. The unit of poor law organisation is the parish, and the poor laws are administered by the parish council. The control of the insane is vested primarily in the Commissioners of Lunacy, and for each county there is a Lunacy Board. There are in addition a number of burghs largely independent of the county council. The burghs of Renfrewshire are Renfrew, Paisley, Greenock, Port Glasgow, Barrhead, Gourock, Johnstone, and Pollokshaws. Renfrew is the only royal burgh. The burghs are managed by town councils, which administer the property of the burghs, impose the rates necessary for upkeep, and make bye-laws for the regulation of the trade of the town and the conduct of the inhabitants. Town councillors are elected for three years, and one-third of the council retires annually. The councillors elect among themselves magistrates, who besides performing other duties, act as judges in the cases that come before the ordinary police courts.

It must not be forgotten that there is still a considerable amount of overlapping and confusion in the adminis-

Hairst Street and Town Hall, Renfrew

trative divisions, not only of Renfrew but of all the counties of Scotland. The registration county is not the same as the civil county; the ecclesiastical parish differs from the civil parish; the district under municipal authority has no fixed relation to any of these other areas. In 1889 under a Local Government Act the Boundary Commissioners rectified some of the most glaring anomalies, and transferred certain areas from one parish to another, and in certain cases from one county to another. The ecclesiastical divisions, however, in many cases still fail to harmonise with the civil divisions.

Since the Education Act of 1872 the management of education in Scotland has been entrusted mainly to School Boards, of which Renfrew has twenty-one. Education is compulsory for children between the ages of five and fourteen years, and is free to all. Above the Primary schools there are two classes of Higher schools called Intermediate and Secondary. The former schools provide a three years' course, and the latter at least a five years' course of education after the elementary stages. Pupils who have passed through a Secondary school with credit are quite able to go with profit direct to the University.

Secondary and technical education is financed largely by a County Committee, which is empowered to give grants to schools, and to assist pupils by bursaries or otherwise. The pupil teacher system is to a great extent abolished. On passing the examinations after three years in an Intermediate school, young people who wish to become teachers may be accepted as Junior Students. They take the usual curriculum of a Secondary school with some training in the art and science of teaching in addition. They then pass to the Training Colleges as students in full training, where they spend two years, after which they are recognised as certificated teachers, although two years of probation must be passed before final recognition is obtained.

22. The Roll of Honour.

It used to be said that one could not throw a stone in Paisley without striking a poet, and there is more truth in this hyperbole than in most. In fact the whole county is marvellously rich in literary associations; and yet, curiously enough, there is not one name quite in the front rank of literary fame. Instead of one star of the first magnitude in an otherwise empty sky, we have a whole constellation of lesser lights; yet the total literary radiance from Renfrewshire, in spite of the small size of the county, will bear comparison with that from any other shire in Scotland. The greatest names connected with the county are not those of literary men. Two of her sons stand head and shoulders above the rest, Scotland's noblest hero, the patriot Wallace, and her greatest inventor, James Watt. Although attempts have been made to dispute Wallace's birthplace, in all probability he was born at Elderslie near the place that is still pointed out as his home. In all Scottish history his is the grandest figure :

> " The manlyast man, the starkest off persoun
>   Leyffand he was ; and als stud in sic rycht
>   We traist weill God his dedis had in sycht."

The most famous of Scottish warriors in things spiritual was also connected with Renfrewshire. Though born in the east, John Knox was descended from the Knoxes of Ranfurly.

James Watt was born in Greenock and lived there till he was eighteen years of age.   He went to Glasgow

James Watt

in 1754, and found employment in the little shop of a mechanic who called himself an optician.   Here Professor

Anderson handed him the model of the Newcomen engine to mend, and so originated one of the greatest discoveries in the history of the world. Watt's memory is kept green in his native town in many ways. A statue, an institution, a monument, a great dock, all bear his name.

The noble house of Sempill, long connected with the county, illustrates how persistently the golden vein of

Castle Semple, Lochwinnoch

literature outcrops in some families. Robert Sempill, born about 1530, was one of the most noted partizans of the Reforming party, and assisted the cause by writing numerous satires, which are coarse and brutal, but at the same time pithy and clever. A " gude, swyne hogge " was one of the mildest of his terms. Then came three generations of Sempills, James, Robert, and Francis, all of whom won distinction by their writings. Sir James, nicknamed "the Dancer" by John Knox owing to his courtly accomplishments, assisted King James VI in the preparation of the *Basilicon Doron*, and wrote polemical works against the Catholic church. Robert is best remembered by his famous poem, *Habbie Simpson, Piper of Kilbarchan*, in a metre which has been traced back to the troubadours :

> "At clerk-plays, when he wont to come,
> His pipe played trimly to the drum,
> Like bykes o' bees he gart it bum,
>   And tuned his reed,
> Now all our pipers may sing dumb
> Sin' Habbie's deid."

Both Ramsay and Burns were influenced by this poem, and copied the form of it, which became characteristic of Scottish vernacular verse. Robert's son Francis became Sheriff-depute of Renfrewshire, and wrote many poems, although the authorship of some attributed to him is doubtful. *Maggie Lauder* is probably his best known song.

Robert Wodrow, author of the *History of the Sufferings of the Church of Scotland*, became minister of Eastwood parish in 1703. Though not quite free from prejudice or credulousness, his work is of very high value for his period. It is only natural that a parish so much associated with the Covenanters as Eaglesham should have produced a historian of those who suffered in the "killing times." John Howie was a farmer near Eaglesham and came of a family that had undergone persecution for adherence to the Covenant. His *Scots Worthies* was at one time one of the most widely read books in Scotland, and even at the present time when interest in the early struggles of the church is waning, the simple, direct, and vivid style of the book gives it a high place as pure literature. From the same district came another book that enjoyed a popularity almost equal to that of the *Scots Worthies*. This was *The Course of Time*, the author of which, Robert Pollok, was born in Eaglesham parish, and educated at Mearns and Fenwick schools. On Pollok's long poem the dust now rests undisturbed, for it has the one vital fault of literature, it is dull. Whole sections resemble a dreary sermon put into tedious blank verse, and yet true poetic gems can be found if the rubbish be sifted with patience.

John Wilson, the author of *The Clyde*, was born in Lanarkshire, but found a home for his later years in Greenock. Like the river whose beauties he sang, he started life among the Southern Uplands, and came to rest where the Clyde enters the sea. He was schoolmaster in turn at Lesmahagow, Rutherglen, and Greenock. The magistrates and the minister of Greenock, however, despising everything without a money value, stipulated that he should abandon "the profane and unprofitable art of poem-making." The combination of adjectives is typical of the true barbarian. Wilson loyally kept his pledge but remained an embittered man, and henceforth, in his own words, passed "the dreary days of an obscure life, the contempt of shopkeepers and brutish skippers."

Although Thomas Campbell was born in Glasgow, in his youth many of his summers were spent in Renfrewshire. To the end of his life he retained tender memories of his boyhood's haunts beside the pleasant banks of the Cart. Indeed the Cart should be the sacred stream of the shire, for by its banks linger memories, not only of Campbell, but of Tannahill, Christopher North, and Alexander Smith.

John Galt, author of *The Annals of the Parish*, was born in Irvine, but in his eleventh year he went with his parents to Greenock. He was employed in the Custom House there until he determined to make his fortune in London, but he returned to Greenock, broken in health, to spend his last days. There is a fountain in his memory on the Esplanade.

Among the many poets of Paisley Tannahill takes first place. He was born in the town, and there he spent most of his life and wrote most of his songs. His best poetry too deals with the scenes among which his life was passed, the Braes o' Gleniffer, and "the birks o' Stanley-shaw."

In Paisley also, in the High Street, Christopher North was born. His remarkable personality is perhaps more interesting now than much of his writing. He was strongly attached to the surroundings where his boyhood was spent, and described Mearns under the name of "our parish" in his *Recreations*. In the manse of that parish he was educated, and the countryside he describes in an impassioned apostrophe as the "fairest of Scotland's thousand parishes." Alexander Smith is also associated with

Paisley. Although born in Kilmarnock he was brought up in Paisley and Glasgow, and he describes the former town under the name of Greysley in *Alfred Hagart's Household*.

Professor John Wilson
(*Christopher North*)

James Young, the founder of the oil-shale industry of Scotland, had his house in Renfrewshire for a time. James Smith of Jordanhill is known to the geologists of many countries by his work on the arctic shells of the Clyde clay beds.

23. THE CHIEF TOWNS AND VILLAGES OF RENFREWSHIRE.

(The figures in brackets after each name give the population in 1911, and those at the end of each section are references to pages in the text.)

Barrhead (11,387) is situated on low ground near the entrance to the Loch Libo Gap through the volcanic hills of Renfrewshire. It is a manufacturing place with machine-tool works, foundries, and other engineering establishments. In or near the town are several dyeing and bleaching works, and one of the largest sanitary engineering works in Scotland. Originally on the site of the town there were several detached villages, but in the last hundred years Barrhead has grown so rapidly as to swallow up these places. Near the town is Boylestone Quarry, a quarry in the volcanic rocks from which native copper is sometimes obtained.

The town has communication by electric tram with Paisley, and with Rouken Glen on the outskirts of Glasgow. On the hills near the town are the reservoirs that supply water to part of the south side of Glasgow. The town itself is not prepossessing in appearance, but some of the valley and moorland scenery in its vicinity is particularly charming. (pp. 18, 25, 27, 35, 37, 41, 56, 80, 84, 89, 141, 147.)

Bishopton is a village situated five miles north-west of Paisley, and is in an agricultural district. At Bishopton the

Caledonian railway enters a deep rock cutting and thence tunnels through a ridge between the village and the Clyde. The making of the tunnel was a matter of great difficulty; it is said that for blasting alone the gunpowder cost £12,000. (pp. 41, 61, 143.)

Bridge of Weir (2716) stands on the river Gryfe three miles north-west of Johnstone. It was one of the earliest cotton-spinning centres of Scotland, but is now chiefly a residential place

Quarrier's Orphan Homes

for people engaged in business in the neighbouring large towns. Near the village stand the ruins of Ranfurly Castle, and not far away is a mound said to mark a Roman fortification. In the vicinity are the well-known Quarrier's Orphan Homes, which accommodate over a thousand boys and girls. (p. 143.)

Busby (1058) lies on the borders of Lanarkshire and Renfrewshire but chiefly in the latter county. It was formerly a very busy manufacturing town with cotton mills, chemical works, and bleach-fields, but many of the mills are now deserted and the population is diminishing. The town is in the neighbourhood of much of the prettiest scenery of the White Cart. It is connected with Glasgow by the branch of the Caledonian Railway that goes to East Kilbride. (p. 22.)

Cathcart (15,205) is now little more than a suburb of Glasgow. It is mainly a residential place, but a few industries such as engineering and paper-making are carried on. The places takes its name from the old castle on the Cart (caer = a castle) the origin of which is lost in antiquity. In the days of Wallace the castle was a place of considerable strength. (pp. 11, 105, 113, 131.)

Clarkston lies a mile nearer Glasgow than Busby. It is the centre of a pleasant, undulating, and well-wooded countryside, and in recent years has grown very rapidly in favour as a residential suburb of Glasgow.

Crosslee is a little village two and a half miles north-west of Johnstone. It was formerly a manufacturing centre, but its industrial life is now extinct. It has entered on a new lease of prosperity as a residential centre.

Eaglesham (1138) is in some ways the most interesting upland village in Renfrewshire. It stands over 500 feet above sea-level at the edge of wide moors that stretch far south into Ayrshire. The moors were the refuge of the Covenanters in the " killing times," and many memories of the martyrs linger round Eaglesham. The present village was founded by the twelfth Earl of Eglinton in 1796, but succeeded an older village which was granted a weekly market in the reign of Charles II. The elevated situation, the fresh moorland air, and the extensive prospects on all sides, have made the place a favourite suburban retreat for city toilers. The absence of railway communication gives the village an added charm. Near it is Pulnoon Castle, said to have been built with the money obtained by Sir Hugh

Montgomery or his father as the ransom of Hotspur, taken prisoner at Otterburn. (pp. 40, 101, 103, 106, 153.)

Elderslie is two miles from Paisley on the Johnstone road. All the associations of the place are linked with the name of Wallace. The house in which he was born is still pointed out, and it is probable at least that the site is the same. The estate was granted to Sir Malcolm Wallace, the patriot's father, in the thirteenth century, and remained in the family till 1769. There is a carpet work in the village. (pp. 103, 143, 150.)

Gourock (7442) was originally built on the raised beach surrounding Gourock Bay, but has now crept up the hill side and has extended for two miles along the shore. The situation is particularly beautiful. In the foreground the sheltered waters of the bay afford a haven for yachts and other small craft, while across the blue waters of the firth rise the smooth slopes above Kilcreggan, and in the background the rugged peaks of the Argyllshire mountains. In summer Gourock is one of the favourite resorts of the Glasgow holdiday-maker, strong points in its favour being its accessibility, its natural beauty, and the facilities afforded by it for all kinds of pastime. It is one of Glasgow's main gates to the Firth of Clyde, for from it starts the fleet of speedy river-steamers owned by the Caledonian Railway Company. Since 1889 the town has been connected with Glasgow by rail. In the basalt quarry behind Gourock good specimens of rare minerals are sometimes obtained, while the sandstones of the neighbourhood are so impregnated with copper as to have been worked for that metal. Gourock is one of the most ancient of Renfrewshire towns. In the seventeenth century Gourock was erected by charter into a burgh of barony with a weekly court and market. (pp. 17, 40, 41, 44, 49, 75, 98, 99, 107, 129, 143, 147, 171.)

Greenock (75,140) ranks seventh in size among the towns of Scotland. It is built largely along the raised beach parallel to

Gourock from Lyle Road

the firth, and rises inland in a series of terraces. Greenock first came into importance in the seventeenth century, when by the influence of John Shaw, the principal proprietor, a charter was obtained from Charles I creating the town a burgh of barony. At first the inhabitants were forbidden to engage in foreign trade, a privilege enjoyed only by royal burghs, but in 1670 Sir John Shaw, being high in favour with the king, was enabled to have the restriction removed. At the end of the seventeenth century the population was about a thousand, and until this time the herring fishing had been the most important occupation of the town, but now cargoes of a general nature began to come into the port, and the place rapidly grew in size and importance. It became necessary to improve the harbour accommodation, and after fruitless attempts to obtain help from parliament, the citizens, headed by their far-seeing laird, Sir John Shaw, determined to carry out the work themselves, and in 1710 the new harbour and quays were successfully completed. About the same time a custom house was established at Greenock which was at first subordinate to Port Glasgow, but later the positions were reversed.

At the Jacobite rebellions of the '15 and the '45 Greenock, like practically all the towns of the west, strongly supported the Hanoverian cause. Towards the end of the century active steps were taken to improve the harbour accommodation, an enterprising policy that has been pursued successfully to the present time. A new quay was built and additions were made from time to time until the Custom House Quay had a thousand feet of frontage. In 1805 the East India Harbour was begun, and in 1818 a new dry dock. Still the port continued increasing in size, and in 1850 the fine Victoria Harbour was opened, and twelve years later the Albert Harbour was begun. Then came the Garvel Graving Dock, next the Princes Pier, and finally the magnificent James Watt Dock with a depth at low water of over 30 feet.

Greenock has been an important ship-building centre since the American War. The yards of Caird and Scott are famous

Greenock from the River

throughout the country. Some of the biggest battle-ships afloat have been built by the latter firm. Naturally the allied industries are well represented in the town. Rope-making and sail-making, saw-mills, the making of anchors and cables, all kinds of deck machinery, and every branch of marine engineering—all these industries flourish in the town. Greenock is still the most important sugar-refining town in Scotland, and in spite of continental competition large cargoes of raw sugar are imported every year. Chemical works add to the wealth if not to the amenities of the town. From any elevated point behind Greenock a magnificent view of the firth and the Highland border can be obtained. Famous names associated with the town are James Watt, Highland Mary, John Wilson, John Galt, and John and Edward Caird. (pp. 10, 12, 17, 27, 46, 47, 58, 61, 66, 75, 81, 82, 83, 84, 85, 90, 92, 94, 96, 97, 98, 99, 107, 111, 135, 143, 146, 147, 151, 154, 155.)

Houston or Hugh's Town (1345) derives its name from Sir Hugh de Padvinan, who obtained a grant of lands in that neighbourhood in the twelfth century. The barony was originally called Kilpeter, after a church dedicated to St Peter, and the name is still found in Peter's Burn, which flows through the village, in St Peter's Well, and in Peter's Day, the name of the old fair day in July. The present village was founded towards the end of the eighteenth century, and stands rather higher up the burn than old Houston. A relic of former days, however, is to be seen in the fine old Cross of Houston. This is a pillar nine feet high fixed in a four-stepped pedestal and surmounted by a sun-dial and a globe. The dial is 200 years old but the pillar is very much older. The most notable mansions in the neighbourhood are Barochan House and Gryfe Castle. (pp. 122, 143.)

Howwood or Hollow-wood lies three miles south-west of Johnstone. Bleach-works in the vicinity give employment to a number of the inhabitants. It is an old-fashioned place near some picturesque reaches of the Black Cart. (p. 19.)

Hurlet is three miles south-east of Paisley. Good seams of coal and iron occur in the locality, and the former mineral has been worked near the village for over three hundred years. For more than a hundred and fifty years chemicals have been made in Hurlet, in fact for some time it was the only place

"Roman" Bridge, Inverkip

in Scotland where copper-sulphate was manufactured. In the vicinity are fine exposures of the fossil-bearing Hurlet Limestone, which is the lowest of the Carboniferous Limestone series, and which attains in places the thickness of a hundred feet.

Inverkip (1168), as its name indicates, is situated at the mouth of the Kip about two and a half miles north of Wemyss Bay. Near the village is the mansion of Ardgowan, the seat of the Shaw Stewarts. In the grounds are the ruins of Inverkip

Castle which probably succeeded the old castle that stood on the same site in the days of Bruce. (pp. 33, 52, 53, 129, 138, 143.)

Johnstone (12,045) stands on the right bank of the Black Cart nearly four miles west of Paisley. It is a good example of the towns that have grown owing to the development of the coal-fields of Scotland, for 150 years ago the town did not exist. At one time hand-loom weaving was the principal occupation of the inhabitants, but this is now extinct. There is still some textile

Kilmacolm, from the East

work, but engineering is the chief industry. All branches of this trade are carried on, but there is a specialisation in machine tools. (pp. 18, 19, 79, 80, 82, 142, 144, 147, 158, 159, 160, 164.)

Kilmacolm (6242) took its name from the ancient church dedicated to St Columba. It is situated in Strathgryfe, a little way from the left bank of the river. Its elevated situation and its nearness to the open moors have made it a favourite residential place. (pp. 122, 143.)

Linwood (2055) stands on the left bank of the Black Cart about a mile and a half north-east of Johnstone. It is an industrial village with a cotton-mill and paper-works.

Lochwinnoch (4254) is pleasantly situated at the edge of Castle Semple Loch among grassy slopes, and sheltered by woods. The principal industry is the making of furniture. Alexander Wilson, poet and naturalist, worked in Lochwinnoch as a hand-loom weaver. (pp. 41, 61, 130.)

Neilston (4616) is situated in the Loch Libo Gap two miles south-west of Barrhead. Most of the village lies over four hundred feet above sea-level. Print-works in the neighbourhood engage most of the inhabitants. Near the village is Neilston Pad, a curious flat-topped eminence which forms a landmark for many miles around. Three miles along the valley to the south-west is Loch Libo, overlooked by Corkindale Law, from which an extensive view can be obtained. The village is on the Glasgow and South-Western Railway to Kilmarnock. (pp. 20, 58, 60.)

Nitshill, situated two miles north-east of Barrhead, lies in the midst of coal and limestone bearing strata. Most of the workers are engaged in mines or quarries. In addition there is a chemical factory.

Paisley (84,477) is the largest town in Renfrewshire and the fifth in all Scotland. It is situated mainly on the left bank of the Cart some three miles from its junction with the Clyde. The town is built on flat, low-lying land, but to the south the ground rises steeply in the escarpment of the Gleniffer Braes.

The first undoubted reference to the place is in a charter granted in 1157 to Walter FitzAlan, in which the lands of Passeleth are mentioned. This ground was given by Walter to the monks whom he brought from England, and a monastery was created, which afterwards became the famous Paisley Abbey. A town soon sprang up and prospered greatly by its vicinity to

Near Lochwinnoch

the powerful abbey. At the end of the fifteenth century, it became a burgh of barony with liberty to buy and sell, to have workmen and sellers of goods, "likewise to possess a cross and market for ever." Hitherto Paisley had been subject to the royal burgh of Renfrew, and violent quarrels took place before Paisley's independence was vindicated. It is interesting to note that exactly the same thing took place when Glasgow was trying to shake off

Paisley Grammar School

the authority of Rutherglen. It is strange to think that there was a time when the great towns of Glasgow and Paisley were subject to their now comparatively tiny rivals, far out-distanced in the race for commercial supremacy.

James IV visited Paisley on more than one occasion, and James VI was entertained in the Abbey although it is said that the bailies, thrifty souls, begged him not to enter the town proper, as the civic purse was too empty to give him a suitable reception.

11—5

The town, like most of the south-west, was strongly anti-Stewart, although Paisley had not suffered so much as some other places during the Covenanting persecutions. The revolution of 1688 was warmly supported, and the town gave active help against the Pretenders in 1715 and 1745, in consequence of which Prince Charles Edward fined the townspeople a thousand pounds, half of

Burgh Hall, Pollokshaws

which was paid. The weavers of Paisley were Radicals to a man, and serious riots took place in the town at the time of the Chartist movement. The trade and manufactures of Paisley began to develop rapidly after the Treaty of Union in 1707. At the present time the town is noted chiefly for its thread-mills and its engineering works. Paisley's educational institutions are exceptionally good. The Grammar School dates from the sixteenth century. Famous names associated with the town are Tannahill, Christopher North, Alexander Smith, William Motherwell, and Alexander Wilson. (pp. 19, 22, 57, 58, 61, 66, 67, 73, 75, 77, 78, 79, 80, 82, 84, 98, 100, 102, 103, 105, 106, 107, 108, 110, 111, 113, 114, 118, 130, 132, 134, 137, 142, 143, 144, 145, 146, 147, 150, 155, 156, 157, 160, 165, 166.)

Pollokshaws (12,932) is now continuous with Glasgow although it remains a separate municipality. It is essentially a manufacturing place, the industries including weaving, printing and bleaching, and paper-making. The town is situated on the White Cart, and even yet many stretches of the river near at hand are picturesque. In the immediate vicinity is Pollok, the seat of the Stirling Maxwells. (pp. 11, 147.)

Port Glasgow (17,749) was founded in 1668, when Glasgow obtained lands at Newark in order to have an outlet for her manufactures by sea. On this ground "they have built ane very fine harbour, some very good houses, both for dwellings, sallaradges, and warehouses." The town grew rapidly, and in 1710 it was made the chief custom-house port of the Clyde. The deepening of the Clyde dealt a staggering blow to the rising port, for traffic now went direct to Glasgow. With praiseworthy energy the inhabitants devoted themselves to ship-building and allied trades, and now Port Glasgow is the greatest ship-building town on the Clyde. Port Glasgow is connected with Greenock and Gourock by electric tram. (pp. 24, 27, 47, 75, 84, 90, 91, 92, 132, 143, 144, 147, 162.)

Renfrew (12,565) comes into prominence at the time of David I, who made the place a royal burgh, although a formal charter was not obtained until the reign of Robert III. In the vicinity of the town there was formerly a royal castle, the residence of the Stewarts, though no trace of it now remains. Far more lasting than the solid walls of masonry are the names

Port Glasgow and Ben Lomond

associated with the castle, such as "The King's Meadow," "The King's Inch," "The Orchard," and "The Dog Row." There were many royal visits to Renfrew till the fifteenth century when the castle fell out of favour. Until the end of the fourteenth century the town was a formidable rival of Glasgow and Rutherglen in size, trade, and prosperity. The town then gradually declined in importance till at the end of the seventeenth century it was much decayed, it had no foreign commerce, and its home trade was "not worth the nameying." It was !not until the middle of the nineteenth century that Renfrew's prospects brightened. Then it shared in the rapidly growing prosperity of Glasgow and was caught up on the same wave of industrial progress. Now it is one of the most flourishing towns on the Clyde. It is the greatest centre for dredger-building in the world, it has the largest boiler-works in Britain, and negotiations are in progress for the construction of a graving dock which in size will beat all records. The industrial future of Renfrew seems rosy indeed. In history the town and its surroundings are associated with the defeat of Somerled, Lord of the Isles, in 1164, the death of Marjory Bruce and the birth of her child, afterwards Robert II, the contest of "Palm-my-arm," and the capture of Argyll in 1685. (pp. 10, 12, 13, 25, 75, 83, 84, 91, 92, 97, 100, 103, 107, 110, 114, 135, 147.)

Wemyss Bay is situated on the coast just on the border of Renfrew and Ayr. It is one of the chief centres of the Clyde tourist traffic, for the Caledonian Railway trains run on to the pier, whence passengers are conveyed by the company's steamers to various places on the firth, chiefly in Bute and Cumbrae. The handsome villas of Wemyss Bay are built mainly of the red sandstones so well developed in the locality. Castle Wemyss is the residence of Lord Inverclyde. (pp. 12, 33, 38, 53, 143, 165.)

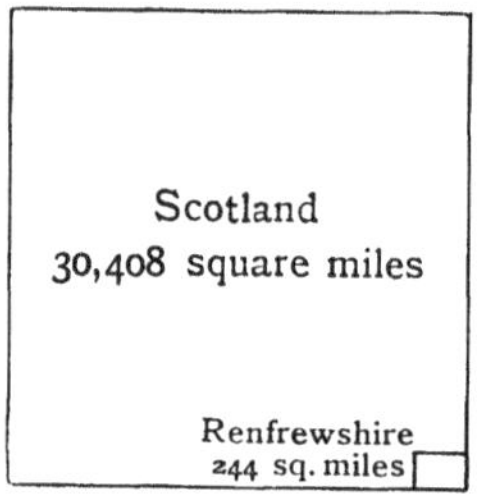

Scotland
30,408 square miles

Renfrewshire
244 sq. miles

Fig. 1. Comparative areas of Scotland and Renfrewshire

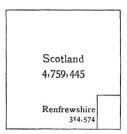

Scotland
4,759,445

Renfrewshire
314,574

Fig. 2. Comparative populations of Scotland and
Renfrewshire, 1911

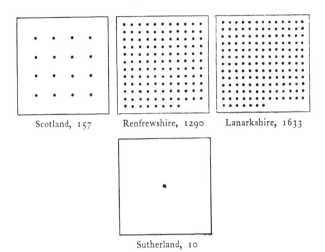

Scotland, 157   Renfrewshire, 1290   Lanarkshire, 1633

Sutherland, 10

Fig. 3. Comparative density of population in Scotland, Renfrewshire, Lanarkshire and Sutherland

(*Each dot represents* 10 *persons*)

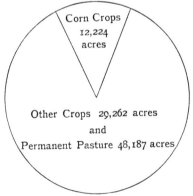

Corn Crops
12,224
acres

Other Crops 29,262 acres
and
Permanent Pasture 48,187 acres

Fig. 4. Proportionate areas of corn and other crops in Renfrewshire, 1910

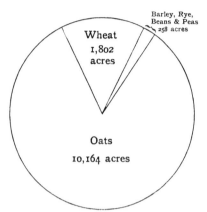

Fig. 5. Proportionate areas of oats, wheat and other crops in Renfrewshire, 1910

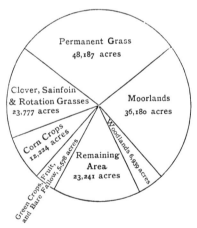

Fig. 6. Proportion of permanent pasture to other areas in Renfrewshire, 1910

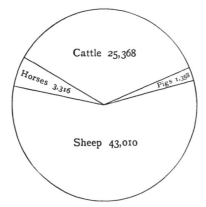

Fig. 7. Proportionate numbers of sheep, cattle, horses
and pigs in Renfrewshire, 1910

www.ingramcontent.com/pod-product-compliance
Ingram Content Group UK Ltd.
Pitfield, Milton Keynes, MK11 3LW, UK
UKHW042144280225
455719UK00001B/76

9 781107 616509